NATEF Standards Job Sheets

Suspension and Steering (A4)

Second Edition

Jack Erjavec

THOMSON

DELMAR LEARNING

Australia   Canada   Mexico   Singapore   Spain   United Kingdom   United States

THOMSON

DELMAR LEARNING

NATEF Standards Job Sheets

Suspension and Steering (A4)
Second Edition

Jack Erjavec

Vice President, Technology and Trades SBU:
Alar Elken

Editorial Editor:
Sandy Clark

Senior Acquisitions Editor:
David Boelio

Development Editor:
Matthew Thouin

Marketing Director:
David Garza

Channel Manager:
William Lawrensen

Marketing Coordinator:
Mark Pierro

Production Director:
Mary Ellen Black

Production Editor:
Toni Hansen

Art/Design Specialist:
Cheri Plasse

Technology Project Manager:
Kevin Smith

Editorial Assistant:
Andrea Domkowski

ISBN-13: 978-1-4180-2077-4
ISBN-10: 1-4180-2077-X

NOTICE TO THE READER

Publisher does not warrant or guarantee any of the products described herein or perform any independent analysis in connection with any of the product information contained herein. Publisher does not assume, and expressly disclaims, any obligation to obtain and include information other than that provided to it by the manufacturer.

The reader is expressly warned to consider and adopt all safety precautions that might be indicated by the activities herein and to avoid all potential hazards. By following the instructions contained herein, the reader willingly assumes all risks in connection with such instructions.

The publisher makes no representation or warranties of any kind, including but not limited to, the warranties of fitness for particular purpose or merchantability, nor are any such representations implied with respect to the material set forth herein, and the publisher takes no responsibility with respect to such material. The publisher shall not be liable for any special, consequential, or exemplary damages resulting, in whole or part, from the readers' use of, or reliance upon, this material.

CONTENTS

PREFACE

The automotive service industry continues to change with the technological changes made by automobile and tool and equipment manufacturers. Today's automotive technician must have a thorough knowledge of automotive systems and components, good computer skills, exceptional communication skills, good reasoning, the ability to read and follow instructions, and above average mechanical aptitude and manual dexterity.

This new edition, like the last, was designed to give students a chance to develop the same skills and gain the same knowledge that today's successful technician has. This edition also reflects the changes in the guidelines established by the National Automotive Technicians Education Foundation (NATEF), as of July 2005.

The purpose of NATEF is to evaluate technician training programs against standards developed by the automotive industry and recommend qualifying programs for certification (accreditation) by ASE (National Institute for Automotive Service Excellence). Programs can earn ASE certification upon the recommendation of NATEF. NATEF's national standards reflect the skills that students must master. ASE certification through NATEF evaluation ensures that certified training programs meet or exceed industry-recognized, uniform standards of excellence.

At the expense of much time and many minds, NATEF has assembled a list of basic tasks for each of their certification areas. These tasks identify the basic skills and knowledge levels that competent technicians have. The tasks also identify what is required for a student to start a successful career as a technician.

Most of the content in this book are job sheets. These job sheets relate to the tasks specified by NATEF. The main considerations during the creation of these job sheets were student learning and program certification by NATEF.

Students are guided through standard industry accepted procedures. While they are progressing, they are asked to report their findings as well as offer their thoughts on the steps they have just completed. The questions asked of the students are thought provoking and require students to apply what they know to what they observe.

The job sheets were also designed to be generic. That is, whenever possible, the tasks can be performed on any vehicle from any manufacturer. Also, completion of the sheets does not require the use of specific brands of tools and equipment; rather students use what is available. In addition, the job sheets can be used as a supplement to any good textbook.

Also included are description and basic use of the tools and equipment listed in NATEF's standards. The standards recognize that not all programs have the same needs, nor do all programs teach all of the NATEF tasks. Therefore, the basic philosophy for the tools and equipment requirement is that the training should be as thorough as possible with the tools and equipment necessary for those tasks.

Theory instruction and hands-on experience of the basic tasks provide initial training for employment in automotive service or further training in any or all of the specialty areas. Competency in the tasks indicates to employers that you are skilled in that area. You need to know the appropriate theory, safety, and support information for each required task. This should include identification and use of the required tools and testing and measurement equipment required for the tasks, the use of current reference and training materials, the proper way to write work orders and warranty reports, and the storage, handling, and use of Hazardous Materials as required by the 'Right to Know Law', and federal, state, and local governments.

Words to the Instructor: I suggest you grade these job sheets based on completion and reasoning. Make sure the students answer all questions. Then look at their reasoning to see if the task was actually completed and to get a feel for their understanding of the topic. It will be easy for students to copy others' measurements and findings, but each student should have their own base of understanding and that will be reflected in their explanations.

Words to the Student: While completing the job sheets, you have a chance to develop the skills you need to be successful. When asked for your thoughts or opinions, think about what you observed. Think about what could have caused those results or conditions. You are not being asked to give accurate explanations for everything you do or observe. You are only asked to think. Thinking leads to understanding. Good technicians are good because they have a basic understanding of what they are doing and of why they are doing it.

Jack Erjavec

SUSPENSION AND STEERING SYSTEMS

To prepare you to learn what you should learn from completing the job sheets, some basics must be covered. This discussion begins with an overview of suspension and steering systems. Emphasis is placed on what they do and how they work, including the major components and designs of suspension and steering systems and their role in the operation of suspension and steering systems of all designs.

Preparing to work on an automobile would not be complete without addressing certain safety issues. This discussion covers what you should and should not do while working on suspension and steering systems, including the proper ways to deal with hazardous and toxic materials.

NATEF's task list for Suspension and Steering Systems certification is also given, with definitions of some terms used to describe the tasks. This list gives you a good look at what the experts say you need to know before you can be considered competent to work on suspension and steering systems.

Following the task list are descriptions of the various tools and types of equipment you need to be familiar with. These are the tools you will use to complete the job sheets. They are also the tools NATEF has identified as being necessary for servicing suspension and steering systems.

After the tool discussion is a cross-reference guide that shows which NATEF tasks are related to specific job sheets. In most cases there is a single job sheet for each task. Some tasks are part of a procedure, in which case one job sheet may cover two or more tasks. The remainder of the book contains the job sheets.

BASIC SUSPENSION AND STEERING SYSTEM THEORY

A vehicle's tires, wheels, and suspension and steering systems provide the contact between the driver and the road. They allow the driver to maneuver the vehicle safely in all types of conditions and to ride in comfort and security.

Suspension and steering systems (Figure 1) have been developed to provide good handling for lighter vehicles and carlike comfort in pickups and SUVs. Computer technology has made possible such features as variable ride qualities and active and adaptive suspensions. These computer systems are integrated with power train controls to ensure improved handling and safety. Systems such as stability control are also linked to the brake system through a computer.

Figure 1 Steering and suspension systems.

Wheels

Wheels are made of either stamped or pressed steel discs riveted or welded together. They are

1

also available in the form of aluminum or magnesium rims that are die-cast or forged. Near the center of the wheel are mounting holes that are tapered to fit lug nuts that center the wheel over the hub. The rim has a hole for the tire's valve stem and a drop center area designed to allow for easy tire removal and installation. Wheel offset is the vertical distance between the centerline of the rim and the mounting face of the wheel.

Wheel size is designated by rim width and rim diameter. Rim width is determined by measuring across the rim between the flanges. Rim diameter is measured across the bead seating areas from the top to the bottom of the wheel.

Tires

The primary purpose of tires is to provide traction. Tires also help the suspension absorb road shocks. They are designed to carry the weight of the vehicle, to withstand side thrust over varying speeds and conditions, and to transfer braking and driving torque to the road.

The cord body or casing of a tire consists of layers of rubber-impregnated cords, called *plies*, which are bonded into a solid unit. The plies determine a tire's strength, handling, ride, amount of road noise, traction, and resistance to fatigue, heat, and bruises. The bead is the portion of the tire that helps keep it in contact with the rim of the wheel. It also provides the air seal on tubeless tires. The bead is constructed of a heavy band of steel wire wrapped into the inner circumference of the tire's ply structure. The tread, or crown, is the portion of the tire that comes in contact with the road surface. It is a pattern of grooves and ribs that provides traction. The grooves are designed to drain off water, and the ribs grip the road surface. Tread thickness varies with tire quality. On some tires, small cuts called *sipes*, are molded into the ribs of the tread. These sipes open as the tire flexes on the road, offering additional gripping action, especially on wet road surfaces. The sidewalls are the sides of the tire's body. They are constructed of thinner material than the tread to offer greater flexibility.

Radial ply tires have body cords that extend from bead to bead at an angle of about 90 degrees—"radial" to the circumferential centerline of the tire—plus two or more layers of relatively inflexible belts under the tread. The belts restrict tread motion during contact with the road, thus improving tread life and traction. Radial ply

tires also offer greater fuel economy, increased skid resistance, and more positive braking.

There are basically three categories of tread patterns: directional, nondirectional, and symmetric or asymmetric. A directional tire is mounted so that it revolves in a particular direction. These tires have an arrow on the sidewalls that show the designed direction of travel. A directional tire performs well only when it is rotating in the direction it was designed to rotate in. A nondirectional tire has the same handling qualities in either direction of rotation. A symmetric tire has the same tread pattern on both sides of the tire. An asymmetric tire has a tread design that is different from one side to the other. Asymmetric tires are typically designed to provide good grip when traveling straight (the inside half) and good grip in turns (the outside half of the tread). Most asymmetric tires are also directional tires.

Tires are classified by their maximum speed and profile (aspect) ratio, size, and load range. A tire's profile is the relation of its cross-section height (from tread to bead) compared to the cross-section width (from sidewall to sidewall).

A properly inflated tire gives the best tire life, riding comfort, handling stability, and fuel economy for normal driving conditions. Too little air pressure can result in tire squeal, hard steering, excessive tire heat, abnormal tire wear, and increased fuel consumption (by as much as 10%). Conversely, an overinflated tire can cause a hard ride, tire bruising, and rapid wear at the center of the tire.

Tire Pressure Monitor System (TPMS)

Some late-model vehicles are equipped with TPMS. This system alerts the driver if one or more tires have low pressure. The system is basically comprised of a tire pressure monitor in each tire and an electronic control unit. The tire pressure monitor valve is a tire pressure sensor and transmitter combined as single unit with tire pressure valve. It measures tire pressure and temperature and transmits ID number for measurement value and identification. A battery is built into the valve. The electronic control unit receives radio signals from the individual tire pressure monitors. If the control unit identifies the signal as being from the vehicle and if the inflation value is equal to or lower than a specified value, it transmits a signal to tire pressure warning lamps.

Wheel Bearings

The purpose of all bearings is to allow a shaft to rotate smoothly in a housing or to allow the housing to rotate smoothly around a shaft. Typically, the bearings on axles that drives the wheels are called axle bearings. The wheel is mounted to the hub of an axle shaft and the shaft rotates within a housing. Wheel bearings are used on nondriving axles. The wheel's hub typically rotates on a shaft called the spindle. Axle bearings are typically serviced with the drive axle. Wheel bearings, however, require periodic maintenance and are often serviced with suspension and brake work.

Often the front wheel hub bearing assembly for driven and nondriven wheels consists of two tapered bearings facing each other. Each of the bearings rides in its own race. Some front wheel bearings are sealed units and are lubricated for life.

Suspension Systems

Like the rest of the systems on cars and light trucks, the suspension system has greatly changed through the years. Improved technology, as well as the quest for great handling and comfortable vehicles, has brought about these changes. Electronic technology has enabled manufacturers to equip many of their vehicles with features such as variable ride control, automatic leveling, active suspension, and adaptive suspension systems.

Suspension systems perform a very complicated function. They must keep the vehicle's wheels lined up with the direction of travel, limit the movement of the vehicle's body during cornering and when going over bumps, and provide a smooth and comfortable ride.

The main component of any suspension system is the spring. Many different spring designs are found on today's vehicles. These spring designs can be grouped into three main categories: coil, leaf, and torsion bar.

Coil springs are the most commonly used spring. They are used in front suspensions with upper and lower control arms and with struts. In rear suspensions, coil springs may be fitted between the rear axle and the body or frame or may be part of a strut assembly.

Leaf springs are normally found on the rear of SUVs and pickup trucks, although some of these vehicles use coil springs. Leaf springs are not used in the front suspension of today's vehicles. Some late-model pickups and SUVs offer air suspension systems. These systems are added to existing leaf-spring suspensions. The air spring is positioned between the center of the leaf spring and the frame of the truck. The air spring serves as an adjustable and additional spring at each end of the axle.

Some vehicles have a single steel leaf spring called a mono-leaf spring. The single leaf is thicker in the center and becomes gradually thinner toward the outer ends. The mono-leaf spring may be mounted longitudinally or transversely and is used in front or rear suspensions.

Many late-model pickups and SUVs use torsion bars in their front suspensions. Torsion bars are primarily used in this type of vehicle because they can be mounted low and out of the way of the driveline components.

Shock Absorbers

Shock absorbers control spring action and oscillations to provide the desired ride quality. They also prevent body sway and lean while cornering and they reduce the tendency of a tire to lift off the road, which improves tire life, traction, and directional stability.

The lower half of a shock absorber is a twin-tube steel unit filled with hydraulic oil and nitrogen gas. In some shock absorbers, the nitrogen gas is omitted. A relief valve is located in the bottom of the unit, and a circular lower mounting is attached to the lower tube. This mounting contains a rubber isolating bushing, or grommets. A piston and rod assembly is connected to the upper half of the shock absorber. This upper portion of the shock absorber has a dust shield that surrounds the lower twin-tube unit. The piston is precision fit in the inner cylinder of the lower unit. A piston rod guide and seal are located in the top of the lower unit. A circular upper mounting with a rubber bushing is attached to the top of the shock absorber.

Strut Design, Front Suspension

A strut-type front suspension is used on most front-wheel-drive cars and some rear-wheel-drive cars. Internal strut design is very similar to shock absorber design, and struts perform the same functions as shock absorbers. In many strut-type suspension systems, the coil spring is mounted on the strut. The coil spring is largely responsible for proper curb riding height.

The lower end of the front suspension strut is bolted to the steering knuckle. An upper strut mount is attached to the strut and is bolted into the chassis strut tower. A lower insulator is positioned between the coil spring and the spring seat on the strut. Another spring insulator is located between the coil spring and the upper strut mount. The two insulators prevent metal-to-metal contact between the spring and the strut, or mount.

A rubber spring bumper is positioned around the strut piston rod to provide cushioning action between the top of the strut and the upper support. The upper strut mount contains a bearing, upper spring seat, and jounce bumper.

When the front wheels are turned, the front strut and coil spring rotate with the steering knuckle on the upper strut mount bearing.

Some cars have a multilink front suspension with a link going from the chassis to the steering knuckle. A bearing is mounted between the link and the steering knuckle, and the wheel and knuckle turn on this bearing and the lower ball joint. Therefore, the coil spring and strut do not turn when the front wheels are turned, and a bearing in the upper strut mount is not required.

MacPherson Struts

The purpose of the main components in a MacPherson strut suspension system may be summarized as follows:

- *Coil springs* allow proper setting of suspension ride heights and control suspension travel during driving maneuvers.

- *Shock absorber struts* provide necessary suspension damping and limit downward wheel movement.

- *Lower control arm* controls lateral (side-to-side) movement of each front wheel.

- *Strut upper mount* insulates the strut and spring from the body and provides a bearing pivot for the strut and spring assembly.

- *Ball joint* connects the outer end of the lower control arm to the steering knuckle and acts as a pivot for the strut, spring, and knuckle assembly.

- *Stabilizer bar* reduces body roll.

Multilink Suspension System

In a multilink front suspension, a short upper link is attached to the chassis and the outer end of this link is connected to a third link. The lower end of the third link is connected through a heavy pivot bearing to the steering knuckle. A rubber insulating bushing connects the inner end of the lower link to the front cross member, and a ball joint connects the outer end of the lower link to the steering knuckle.

The shock absorbers are connected from the lower end of the third link to a reinforced area of the fender. A coil spring seat is attached to the lower end of the shock absorber, and the upper spring seat is located on the upper shock absorber mounting insulator. Since the steering knuckle pivots on the lower ball joint and the upper pivot bearing, the coil spring and shock absorber do not rotate with the knuckle. Tension or strut rods are connected from the lower links to tension rod brackets attached to the chassis. A stabilizer bar is mounted on rubber insulating bushings in the tension rod brackets, and the outer ends of this bar are attached to the third link.

Double Wishbone Front Suspension System

In double wishbone suspension systems, the upper and lower control arms have a wishbone shape. Positioning the ball joints and steering knuckle inside the wheel profile increases suspension rigidity. The double wishbones are attached to the chassis, and the upper and lower control arms are attached through bushings to a compliance pivot assembly. When one of the front wheels is subjected to rearward force by hard braking or a road irregularity, the coil spring is compressed and the ride height is lowered. This rearward force on the front wheel twists the compliance pivot, allowing both control arms to pivot slightly. Under this condition, the upper and lower control arm movement allows the front wheel to move rearward a small amount, and this wheel movement absorbs energy to improve ride quality significantly. While cornering, the compliance pivot does not move.

Some multilink front suspensions have compression and lateral lower arms. The lateral arm prevents front wheel movement, and the compression arm prevents fore-and-aft front wheel movement. A ball joint in the outer end of the lateral arm is bolted into the steering knuckle. A rubber insulating bushing in the inner end of the compression arm is bolted to the chassis, and a ball joint in the outer end of this arm is bolted into the steering knuckle.

Strut Design, Rear Suspension

In some rear suspension systems, the lower end of the strut is bolted to the spindle, and the top of the strut is connected through a strut mount to the chassis. The rear coil springs are mounted separately from the struts. These springs are mounted between the lower control arms and the chassis.

In other rear suspension systems, the coil springs are mounted on the rear struts. An upper insulator is positioned between the top of the spring and the upper spring support, and a lower insulator is located between the bottom of the spring and the spring mount on the strut. A rubber spring bumper is positioned on the strut piston rod.

Ball Joints

Ball joints act as pivot points that allow the front wheels and spindles or knuckles to turn between the upper and lower control arms. Ball joints may be grouped into two classifications, load-carrying and nonload-carrying. The coil spring is seated on the control arm to which the load-carrying ball joint is attached. In a torsion bar suspension, the load-carrying ball joint is mounted on the control arm to which the torsion bar is attached. A load-carrying ball joint supports the vehicle weight.

When the lower control arm is positioned below the steering knuckle, vehicle weight pulls the ball joint away from the knuckle. This type of ball joint mounting is referred to as tension loaded and is mounted in the lower control arm with the ball joint stud facing upward into the knuckle.

Many load-carrying ball joints have built-in wear indicators; an indicator on the grease nipple surface recedes into the housing as the joint wears.

A nonload-carrying ball joint may be referred to as a *stabilizing* or *follower* ball joint. A nonload-carrying ball joint is designed with a preload, which provides damping action.

Stabilizer Bar

The stabilizer bar is attached to the cross member and interconnects the lower control arms. Rubber insulating bushings are used at all stabilizer bar attachment points. When jounce and rebound wheel movements affect one front wheel, the stabilizer bar transmits part of this movement to the opposite lower control arm and wheel, which reduces and stabilizes body roll.

Strut Rod

On some front suspension systems, a strut rod is connected from the lower control arm to the chassis. The strut rod is bolted to the control arm, and a large rubber bushing surrounds the strut rod in the chassis opening. The outer end of the strut rod is threaded, and steel washers are positioned on each side of the strut rod bushing. The strut rod prevents fore-and-aft movement of the lower control arm.

Short-and-Long-Arm Front Suspension Systems

A short-and-long-arm (SLA) front suspension system has coil springs with upper and lower control arms. The upper control arm is shorter than the lower control arm. During wheel jounce and rebound travel in this suspension system, the upper control arm moves in a shorter arc than the lower control arm. This action moves the top of the tire in and out slightly, but the bottom of the tire remains in a more constant position.

The inner end of the lower control arm contains large rubber insulating bushings, and the ball joint is attached to the outer end of the control arm. The lower control arm is bolted to the front cross member, and the attaching bolts are positioned in the center of the lower control arm bushings. A spring seat is located in the lower control arm. An upper control arm shaft is bolted to the frame, and rubber insulators are located between this shaft and the control arm.

On some SLA front suspension systems, the coil spring is positioned between the upper control arm and the chassis.

Steering Knuckle

The upper and lower ball joints connect to the steering knuckle. The wheel hub and bearings are positioned on the steering knuckle extension, and the wheel assembly is bolted to the wheel hub. When the steering wheel is turned, the steering gear and linkage turn the steering knuckle. During this turning action, the steering knuckle pivots on the upper and lower ball joints.

Manual Steering Systems

The steering system is composed of three major subsystems: the steering linkage, steering gear, and steering column and wheel. As the driver turns the steering wheel, the steering gear transfers this motion to the steering linkage. The steering linkage turns the wheels to control the vehicle's direction. Although there are many variations to this system, these three major assemblies are in all steering systems.

The term *steering linkage* is applied to the system of pivots and connecting parts that is placed between the steering gear and the steering arms attached to the front or rear wheels, controlling the direction of vehicle travel. The steering linkage transfers the motion of the steering gear output shaft to the steering arms, turning the wheels to maneuver the vehicle.

Parallelogram Steering Linkages

In a parallelogram steering linkage, the tie-rods have ball socket assemblies at each end. One end is attached to the wheel's steering arm and the other end to the center link.

The components in a parallelogram steering linkage arrangement are the pitman arm, idler arm, links, and tie rods. The pitman arm connects the linkage to the steering column through a steering gear located at the base of the column. It transmits the motion it receives from the gear to the linkage, causing the linkage to move left or right to turn the wheels in the appropriate direction. It also serves to maintain the height of the center link. This ensures that the tie rods are parallel to the control arm movement and can avoid unsteady toe settings. *Toe* is a term that defines how well the tires point to the direction of the vehicle.

The idler arm is normally attached to the pitman arm and the frame, supporting the center link at the correct height. A pivot built into the arm or assembly permits sideways movement of the linkage. On some linkages, such as those on a few light-duty trucks, two idler arms are used.

Links can be referred to as center, drag, or steering links. Their purpose is to control sideways linkage movement, which changes the wheel directions. Because they usually are also mounting locations for tie rods, they are very important for maintaining correct toe settings. Center links and drag links can be used either alone or in conjunction with each other, depending on the particular steering design.

Tie rods are the assemblies that make the final connection between the steering linkage and steering knuckles. They consist of inner tie-rod ends, which are connected to the opposite sides of the center link; outer tie-rod ends, which connect to the steering knuckles; and adjusting sleeves or bolts, which join the inner and outer tie-rod ends, permitting the tie-rod length to be adjusted for correct toe settings.

Rack-and-Pinion Steering Linkage

Rack-and-pinion steering has fewer components than parallelogram steering. Tie rods are used in the same fashion on both systems, but the resemblance stops there. Steering input is received from a pinion gear attached to the steering column. This gear moves a toothed rack that is attached to the tie rods.

In the rack-and-pinion steering arrangement, there is no pitman arm, idler arm assembly, or center link. The rack performs the task of the center link. Its movement pushes and pulls the tie rods to change the wheel's direction. The tie rods are the only steering linkage parts used in a rack-and-pinion system.

Recirculating Ball Steering

The recirculating ball is generally found in larger cars or trucks. A sector shaft is supported by needle bearings in the housing and a bushing in the sector cover. A ball nut with threads that mate to the threads of the worm shaft is used, steering wheel movement is transferred via continuous rows of ball bearings between the sector and worm shafts. Ball bearings recirculate through two outside loops, referred to as *ball return guide tubes*. The ball nut has gear teeth cut on one face that mesh with gear teeth on the sector shaft. As the steering wheel is rotated, the worm shaft rotates, causing the ball nut to move up or down the worm shaft. Since the gear teeth on the ball nut are meshed with the gear teeth on the sector shaft, the movement of the nut causes the sector shaft to rotate and swing the pitman arm.

The design of two separate circuits of balls results in an almost friction-free operation of the ball nut and the worm shaft. When the steering wheel is turned, the ball bearings roll in the ball thread grooves of the worm shaft and ball nut. When the ball bearings reach the end of their respective circuits, they enter the guide tubes and are returned to the other end of the circuits.

The teeth on the sector shaft and the ball nut are designed so that an interference fit exists between the two when the front wheels are straight ahead. This interference fit eliminates gear tooth lash for a positive feel when driving straight ahead. An adjusting screw that moves the sector shaft axially obtains proper engagement of the sector and ball nut.

The worm thrust bearing adjuster can be turned to provide proper preloading of the worm thrust bearings. Worm bearing preload eliminates worm endplay and is necessary to prevent steering free-play and vehicle wander.

Worm and Roller Steering

The worm and roller gearbox is similar to the recirculating ball except that a single roller replaces the balls and ball nut. This reduces internal friction, making it ideal for smaller cars. The steering linkage used with a worm and roller gearbox typically includes a pitman arm, center link, idler arm, and two tie-rod assemblies. The function of these components is the same as the parallelogram steering linkage earlier in this chapter.

Steering Wheel and Column

The purpose of the steering wheel and column is to produce the necessary force to turn the steering gear. The exact type of steering wheel and column depends on the year and the car manufacturer. The steering column, also called a steering shaft, relays the movement of the steering wheel to the steering gear. Differences in steering wheel and column designs include fixed column, telescoping column, tilt column, manual transmission, floor shift, and automatic transmission column shift.

Steering Damper

The purpose of a steering damper is simply to reduce the amount of road shock that is transmitted up through the steering column. Steering dampers are mostly found on four-wheel-drive (4WD) vehicles, especially those fitted with large tires. The damper serves the same function as a shock absorber but is mounted horizontally to the steering linkage—one end to the center link, the other to the frame.

Power Steering

The power steering unit is designed to reduce the amount of effort required to turn the steering wheel. It also reduces driver fatigue on long trips and makes it easier to steer the vehicle at slow road speeds, particularly during parking.

Power steering can be broken down into two design arrangements: conventional and nonconventional or electronically controlled. In the conventional arrangement, hydraulic power is used to assist the driver. In the nonconventional arrangement, an electric motor and electronic controls provide power assistance in steering.

There are several power steering systems in use on passenger cars and light-duty trucks. The most common ones are the external piston linkage, integral piston, and power-assisted rack-and-pinion system.

Several of the manual steering parts, such as the steering linkage, are used in conventional power steering systems. The components that have been added for power steering provide the hydraulic power that drives the system. They include the power steering pump, flow control and pressure relief valves, reservoir, spool valves and power pistons, hydraulic hose lines, and gearbox or assist assembly on the linkage.

The power steering pump is used to develop hydraulic flow, which provides the force needed to operate the steering gear. The pump is belt driven from the engine crankshaft, providing flow any time the engine is running. The pump assembly includes a reservoir and an internal flow control valve. The drive pulley is normally pressed onto the pump's shaft.

A pressure relief valve controls the pressure output from the pump. This valve is necessary because of the variations in engine revolutions per minute (rpm) and the need for consistent steering ability in all ranges, from idle to highway speeds. It is positioned in a chamber that is exposed to pump outlet pressure at one end and supply hose pressure at the other. A spring is used at the supply pressure end to help maintain a balance.

A power steering gearbox is basically the same as a manual recirculating ball gearbox with the addition of a hydraulic assist. A power steering gearbox is filled with hydraulic fluid and uses a control valve.

In a power rack-and-pinion gear, the movement of the rack is assisted by hydraulic pressure. When the wheel is turned, the rotary valve changes hydraulic flow to create a pressure differential on either side of the rack. The unequal pressure causes the rack to move toward the lower pressure, thus reducing the effort required to turn the wheels.

Power steering hoses transmit power (fluid under pressure) from the pump to the steering gearbox, ultimately returning the fluid to the pump reservoir. Through material and construction, hoses also function as additional reservoirs and act as sound and vibration dampers.

Electronically Controlled Power Steering Systems

The object of power steering is to make steering easier at low speeds, especially while parking. However, higher steering efforts are desirable at higher speeds in order to provide improved down-the-road feel. The electronically controlled power steering systems provide both of these benefits. The hydraulic boost of these systems is tapered off by electronic control as road speed increases.

When the vehicle is operating at low speeds, the computer supplies a signal to cycle the solenoid faster so that it allows high pump pressure. This provides for maximum power assist during cornering and parking. As the vehicle's speed increases, the solenoid cycles less and the pump provides less assistance. This gives the driver better road feel at high speeds.

Electric/Electronic Rack-and-Pinion System

The electric/electronic rack-and-pinion unit replaces the hydraulic pump, hoses, and fluid associated with conventional power steering systems with electronic controls and an electric motor located concentric to the rack itself. The housing and rack are designed so that the rotary motion of the motor's armature can be transferred to linear movement of the rack through a ball nut with thrust bearings. The armature is mechanically connected to the ball nut through an internal/external spline arrangement.

The basis of system operation is its ability to change the rotational direction of the electric motor while being able to deliver the necessary amount of current to meet torque requirements at the same time. The system monitors steering wheel movement through a sensor mounted on the input shaft of the rack-and-pinion steering gear. After receiving directional and load information from the sensor, an electronic controller activates the motor to provide power assistance.

Unlike conventional power steering, electric/electronic units provide power assistance even when the engine stalls since the power source is the battery rather than the engine-driven pump. The feel of the steering can also be adjusted to match the particular driving characteristics of cars and drivers, from high-performance to luxury touring cars. This design also eliminates hydraulic oil, which means no leaks.

Wheel Alignment

Wheel alignment allows the wheels to roll without scuffing, dragging, or slipping in different types of road conditions. Proper alignment of both the front and the rear wheels ensures greater safety in driving, easier steering, longer tire life, reduction in fuel consumption, and less strain on the parts that make up the steering and suspension systems of the vehicle.

The alignment angles are designed to properly locate the vehicle's weight on moving parts and to facilitate steering. If these angles are not correct, the vehicle is misaligned. The proper alignment of a suspension or steering system centers on the accuracy of the following angles.

- *Caster* is the angle of a wheel's steering axis from the vertical, as viewed from the side of the vehicle. It is the forward or rearward tilt from the vertical line. Caster is designed to provide steering stability.

- *Camber* is the angle represented by the tilt of either the front or rear wheels inward or outward from the vertical, as viewed from the front of the car. Camber is designed into the vehicle to compensate for road crown, passenger weight, and vehicle weight.

- *Toe* is the distance comparison between the leading edge and the trailing edge of the front tires. If the leading edge distance is less, then there is toe in. If it is greater, there is toe out. Toe is critical as a tire-wearing angle.

SAFETY

In an automotive repair shop, there is great potential for serious accidents simply because of the nature of the business and the equipment used. Through carelessness, the automotive repair industry can be one of the most dangerous occupations. The chances of being injured while working on a car are close to nil, however, if you learn to work safely and use common sense. Safety is the responsibility of everyone in the shop.

Personal Protection

Some procedures, such as grinding, result in tiny particles of metal and dust being thrown off at very high speeds. These metal and dirt particles can easily get into your eyes, causing scratches or cuts on your eyeball. Pressurized gases and liquids escaping a ruptured hose or hose fitting can spray a great distance. If these chemicals get into your eyes, they can cause blindness. Dirt and sharp bits of corroded metal can easily fall into your eyes while you are working under a vehicle.

Eye protection should be worn whenever you are exposed to these risks. To be safe, you should wear safety glasses whenever you are working in the shop. Some procedures may require that you wear other eye protection in addition to safety glasses. When cleaning parts with a pressurized spray, for instance, you should wear a face shield. The face shield not only gives added protection to your eyes, but it also protects the rest of your face.

If chemicals such as battery acid, fuel, or solvents get into your eyes, flush them continuously with clean water. Have someone call a doctor, and get medical help immediately.

Your clothing should be well fitted and comfortable but made with strong material. Loose, baggy clothing can easily get caught in moving parts and machinery. Some technicians prefer to wear coveralls or shop coats to protect their personal clothing. Your work clothing should offer you some protection but should not restrict your movement.

Long hair and loose, hanging jewelry can create the same type of hazard as loose-fitting clothing—they can get caught in moving engine parts and machinery. If you have long hair, tie it back or tuck it under a cap.

Never wear rings, watches, bracelets, or neck chains. These can easily get caught in moving parts and cause serious injury.

Always wear leather or similar material shoes or boots with nonslip soles. Steel-tipped safety shoes can give added protection to your feet. Jogging or basketball shoes, street shoes, and sandals are inappropriate in the shop.

Good hand protection is often overlooked. A scrape, cut, or burn can limit your effectiveness at work for many days. A well-fitted pair of heavy work gloves should be worn during operations such as grinding and welding or when handling high-temperature components. Always wear approved rubber gloves when handling strong and dangerous caustic chemicals.

Many technicians wear thin, surgical-type latex gloves whenever they are working on vehicles. These offer little protection against cuts but do offer protection against disease and grease buildup under and around your fingernails. These gloves are comfortable and are quite inexpensive.

Accidents can be prevented simply by the way you act. Following are some guidelines for working in a shop. This list does not include everything you should or shouldn't do; it merely provides some things to think about.

- Never smoke while working on a vehicle or while working with any machine in the shop.

- Playing around is not fun when it sends someone to the hospital.

- To prevent serious burns, keep your skin away from hot metal parts such as the radiator, exhaust manifold, tailpipe, catalytic converter, and muffler.

- Always disconnect electric engine-cooling fans when working around the radiator. Many of these turn on without warning and can easily chop off a finger or hand. Make sure you reconnect the fan after you have completed your repairs.

- When working with a hydraulic press, make sure the pressure is applied in a safe manner. It is generally wise to stand to the side when operating the press.

- Properly store all parts and tools by putting them away in a place where people will not trip over them. This practice not only cuts down on injuries, it also reduces time wasted looking for a misplaced part or tool.

Work Area Safety

Your entire work area should be kept clean and dry. Any oil, coolant, or grease on the floor can make it slippery. To clean up oil, use commercial oil absorbent. Keep all water off the floor. Water is slippery on smooth floors, and electricity flows well through water. Aisles and walkways should be kept clean and wide enough to easily move through. Make sure the work areas around machines are large enough for safe operation of the machine.

Gasoline is a highly flammable volatile liquid. Something that is *flammable* catches fire and burns easily. A *volatile* liquid is one that vaporizes very quickly. *Flammable volatile* liquids are potential firebombs. Always keep gasoline or diesel fuel in an approved safety can and never use gasoline to clean your hands or tools.

Handle all solvents (or any liquids) with care to avoid spillage. Keep all solvent containers closed, except when pouring. Proper ventilation is very important in areas where volatile solvents and chemicals are used. Solvents and other combustible materials must be stored in approved and designated storage cabinets or rooms with adequate ventilation. Never light matches or smoke near flammable solvents and chemicals, including battery acids.

Oily rags should also be stored in an approved metal container. When oily, greasy, or paint-soaked rags are left lying about or are not stored properly, they can combust spontaneously. Spontaneous combustion refers to a fire that starts by itself, without a match.

Disconnecting the vehicle's battery before working on the electrical system, or before welding, can prevent fires caused by a vehicle's electrical system. To disconnect the battery, remove the negative or ground cable from the battery and position it away from the battery.

Know where all the shop's fire extinguishers are located. Fire extinguishers are clearly labeled as to type and types of fire they should be used on. Make sure you use the correct type of extinguisher for the type of fire you are dealing with. A multipurpose dry chemical fire extinguisher puts out ordinary combustibles, flammable liquids, and electrical fires. Never put water on a gasoline fire—the water just spreads the fire. The proper fire extinguisher smothers the flames.

During a fire, never open doors or windows unless it is absolutely necessary; the extra draft only makes the fire worse. Make sure the fire department is contacted before or during your attempt to extinguish a fire.

Air Bag Safety

When service is performed on any air bag system component, always disconnect the negative battery cable, isolate the cable end, and wait for the amount of time specified by the vehicle manufacturer before proceeding with the necessary diagnosis or service. The average waiting period is two minutes, but some vehicle manufacturers specify up to 10 minutes. Failure to observe this precaution may cause accidental air bag deployment and personal injury.

Replacement air bag system parts must have the same part number as the original part. Replacement parts of lesser or questionable quality must not be used. Improper or inferior components may result in inappropriate air bag deployment and injury to the vehicle occupants.

Do not strike or jar a sensor or an air bag system diagnostic monitor (ASDM). This may cause air bag deployment or make the sensor inoperative. Accidental air bag deployment may cause personal injury, and an inoperative sensor may result in air bag deployment failure, causing personal injury to vehicle occupants.

All sensors and mounting brackets must be properly torqued to ensure correct sensor operation before an air bag system is powered up. If sensor fasteners do not have the proper torque, improper air bag deployment may result in injury to vehicle occupants.

When working on the electrical system on an air bag–equipped vehicle, use only the vehicle manufacturer's recommended tools and service procedures. The use of improper tools or service procedures may cause accidental air bag deployment and personal injury. For example, do not use 12V or self-powered test lights when servicing the electrical system on an air bag–equipped vehicle.

Tool and Equipment Safety

Careless use of simple hand tools, such as wrenches, screwdrivers, and hammers causes many shop accidents that could be prevented. Keep all hand tools free of grease and in good condition. Tools that slip can cause cuts and bruises. If a tool slips and falls into a moving part, it can fly out and cause serious injury.

Use the proper tool for the job. Make sure the tool is of professional quality. Using poorly made tools or the wrong tools can damage parts, the tool itself, or you. Never use broken or damaged tools.

Safety around power tools is very important. Serious injury can result from carelessness. Always wear safety glasses when using power tools. If the tool is electrically powered, make sure it is properly grounded. Before using it, check the wiring for bare wires and for cracks in the insulation. When us-ing electrical power tools, never stand on a wet or damp floor. Never leave a running power tool unattended.

Tools that use compressed air are called pneumatic tools. Compressed air is used to inflate tires, apply paint, and drive tools. Compressed air can be dangerous when it is not used properly.

When using compressed air, wear safety glasses or a face shield, or both. Particles of dirt and pieces of metal blown by the high-pressure air can penetrate your skin or get into your eyes.

Before using a compressed air tool, check all hose connections. Always hold an air nozzle or air control device securely when starting or shutting off the compressed air. A loose nozzle can whip suddenly and cause serious injury. Never point an air nozzle at anyone. Never use compressed air to blow dirt from your clothes or hair. Never use compressed air to clean the floor or workbench.

Always be careful when raising a vehicle on a lift or a hoist. Adapters and hoist plates must be positioned correctly to prevent damage to the underbody of the vehicle. There are specific lift points, which allow the weight of the vehicle to be evenly supported by the adapters or hoist plates. The correct lift points can be found in the vehicle's service manual. Before operating any lift or hoist, carefully read the operating manual and follow the operating instructions.

Once you know the lift supports are property-ly positioned under the vehicle, raise the lift until the supports contact the vehicle. Then, check the supports to make sure they are in full contact with the vehicle. Shake the vehicle to make sure it is securely balanced on the lift, and then raise the lift to the desired working height. Before working under a car, make sure the lift's locking devices are engaged.

A vehicle can be raised off the ground by a hydraulic jack. The jack's lifting pad must be posi-tioned under an area of the vehicle's frame or at one of the manufacturer's recommended lift points. Never place the pad under the floor pan or under steering and suspension components, which are easily damaged by the weight of the vehicle. Always position the jack so the wheels of the vehicle can roll as the vehicle is being raised.

Safety stands, also called jack stands, should be placed under a sturdy chassis member, such as the frame or axle housing, to support the vehicle after it has been raised by a jack. Once the safety stands are in position, the hydraulic pressure in the jack should be released slowly until the weight of the vehicle is on the stands. Never move under a vehicle when it is supported only by a hydraulic jack. Rest the vehicle on the safety stands before moving under the vehicle.

Cleaning parts is a necessary step in most repair procedures. Always wear the appropriate protection when using chemical, abrasive, and thermal cleaners.

Vehicle Operation

When a customer brings a vehicle in for service, certain driving rules should be followed to ensure your safety and the safety of those working around you. Before moving a car into the shop, buckle the safety belt. Make sure no one is near, the way is clear, and there are no tools or parts under the car before you start the engine.

Check the brakes before putting the vehicle in gear. Then, drive slowly and carefully in and around the shop.

If the engine must be running while work is done on the car, block the wheels to prevent the car from moving. Put the transmission into park for automatic transmissions or in neutral for manual transmissions. Set the parking (emer-gency) brake. Never stand directly in front of or behind a running vehicle.

Run the engine only in a well-ventilated area to avoid the danger of poisonous carbon monox-ide (CO) in the engine exhaust. CO is an odorless but deadly gas. Most shops have an exhaust ven-tilation system; always use it. Connect the hose from the vehicle's tailpipe to the intake for the vent system. Make sure the vent system is turned on before running the engine. If the work area does not have an exhaust venting system, use a hose to direct the exhaust out of the building.

HAZARDOUS MATERIALS AND WASTES

A typical shop contains many potential health hazards for those working in it. These hazards can cause injury, sickness, impairment, discomfort, and even death. Here is a short list of the different classes of hazards.

- Chemical hazards are caused by high concentrations of vapors, gases, or solids (in the form of dust).

- Hazardous wastes are substances that are the result of a service.

- Physical hazards include excessive noise, vibration, pressures, and temperatures.

- Ergonomic hazards are conditions that impede normal or proper body position and motion.

There are many government agencies charged with ensuring safe work environments for all workers, including the Occupational Safety and Health Administration (OSHA), Mine Safety and Health Administration (MSHA), and National Institute for Occupational Safety and Health (NIOSH). These agencies, in addition to state and local governments, have instituted regulations that must be understood and followed. Everyone in a shop is responsible for adhering to these regulations.

An important part of a safe work environment is the employees' knowledge of potential hazards. Right-to-know laws concerning all chemicals protect every employee in the shop. The general intent of right-to-know laws is for employers to provide their employees with a safe working place as it relates to hazardous materials.

All employees must be trained about their rights under the legislation, the nature of the hazardous chemicals in their workplace, and the contents of the labels on the chemicals. All the information about each chemical must be posted on Material Safety Data Sheets (MSDS) and must be accessible. The manufacturer of the chemical must give these sheets to its customers on request. MSDS detail the chemical composition and precautionary information for all products that can present a health or safety hazard.

Employees must become familiar with the general uses of protective equipment NAT for, accident or spill procedures with, and any other information regarding the safe handling of a hazardous material. This training must be given to employees annually and provided to new employees as part of their job orientation.

All hazardous material must be properly labeled, indicating what health, fire, or reactivity hazard it poses and what protective equipment is necessary when handling each chemical. The manufacturer of the hazardous materials must provide all warnings and precautionary information, which must be read and understood by the user before use. A list of all hazardous materials used in the shop must be posted for the employees to see.

Shops must maintain documentation on the hazardous chemicals in the workplace, proof of training programs, records of accidents or spill incidents, satisfaction of employee requests for specific chemical information via the MSDS, and a general right-to-know compliance procedure manual used in the shop.

When handling any hazardous materials or hazardous waste, make sure you follow the required procedures for handling such material. Also wear the proper safety equipment listed on the MSDS, which includes the use of approved respirator equipment.

Some of the common hazardous materials that automotive technicians use are cleaning chemicals, fuels (gasoline and diesel), paints and thinners, battery electrolyte (acid), used engine oil, refrigerants, and engine coolant (antifreeze).

Many repair and service procedures generate what are known as hazardous wastes. Dirty solvents and cleaners are good examples of hazardous wastes. Something is classified as a hazardous waste if it is on the Environmental Protection Agency (EPA) list of known harmful materials or has one or more of the following characteristics.

- *Ignitability.* A liquid with a flash point below 140°F or a solid that can spontaneously ignite.

- *Corrosivity.* A substance that dissolves metals and other materials or burns the skin.

- *Reactivity.* Any material that reacts violently with water or other materials or releases cyanide gas, hydrogen sulfide gas, or similar gases when exposed to low-pH acid solutions. This includes material that generates toxic mists, fumes, vapors, and flammable gases.

■ *EP toxicity.* Materials that leach one or more of eight heavy metals in concentrations greater than 100 times primary drinking water standard concentrations.

Complete EPA lists of hazardous wastes can be found in the Code of Federal Regulations. It should be noted that no material is considered hazardous waste until the shop is finished using it and ready to dispose of it.

The following list describes the recommended procedures for dealing with some of the common hazardous wastes. Always follow these or any other mandated procedures.

Oil Recycle oil. Set up equipment, such as a drip table or screen table with a used oil collection bucket, to collect oils dripping off parts. Place drip pans underneath vehicles that are leaking fluids. Do not mix other wastes with used oil, except as allowed by your recycler. Used oil generated by a shop (and oil received from household "do-it-yourself" generators) may be burned on site in a commercial space heater. Used oil also may be burned for energy recovery. Contact state and local authorities to determine requirements and to obtain necessary permits.

Oil filters Drain for at least 24 hours, crush, and recycle used oil filters.

Batteries Recycle batteries by sending them to a reclaimer or back to the distributor. Keep shipping receipts to demonstrate that you have recycled. Store batteries in a watertight, acid-resistant container. Inspect batteries for cracks and leaks when they come in. Treat a dropped battery as if it were cracked. Acid residue is hazardous because it is corrosive and may contain lead and other toxics. Neutralize spilled acid by using baking soda or lime, and dispose of as hazardous material.

Metal residue from machining Collect metal filings when machining metal parts. Keep separate and recycle if possible. Prevent metal filings from falling into a storm sewer drain.

Refrigerants Recover or recycle refrigerants (or both) during the service and disposal of motor vehicle air conditioners and refrigeration equipment. It is not allowable to knowingly vent refrigerants to the atmosphere. Recovering and recycling during servicing must be performed by an EPA-certified technician using certified equipment and following specified procedures.

Solvents Replace hazardous chemicals with less toxic alternatives that have equal performance. For example, substitute water-based cleaning solvents for petroleum-based solvent degreasers. To reduce the amount of solvent used when cleaning parts, use a two-stage process (dirty solvent followed by fresh solvent). Hire a hazardous waste management service to clean and recycle solvents. (Some spent solvents must be disposed of as hazardous waste, unless recycled properly.) Store solvents in closed containers to prevent evaporation. Evaporation of solvents contributes to ozone depletion and smog formation. In addition, the residue from evaporation must be treated as a hazardous waste. Properly label spent solvents and store on drip pans or in diked areas and only with compatible materials.

Containers Cap, label, cover, and properly store above ground and outdoors all liquid containers and small tanks within a diked area and on a paved impermeable surface to prevent spills from running into surface or ground water.

Other solids Store materials such as scrap metal, old machine parts, and worn tires under a roof or tarpaulin to protect them from the elements and to prevent potentially contaminated runoff. Consider recycling tires by retreading them.

Liquid recycling Collect and recycle coolants from radiators. Store transmission fluids, brake fluids, and solvents containing chlorinated hydrocarbons separately, and recycle or dispose of them properly.

Shop towels and rags Keep waste towels in a closed container marked "Contaminated Shop Towels Only." To reduce costs and liabilities associated with disposal of used towels, which can be classified as hazardous wastes, investigate using a laundry service that is able to treat the wastewater generated from cleaning the towels.

Waste storage   Always keep hazardous waste separate, properly labeled, and sealed in the recommended containers. The storage area should be covered and may need to be fenced and locked if vandalism could be a problem. Select a licensed hazardous waste hauler after seeking recommendations and reviewing a firm's permits and authorizations.

NATEF TASK LIST FOR SUSPENSION AND STEERING SYSTEMS

A. General Suspension and Steering Systems Diagnosis

A.1.   Complete work order to include customer information, vehicle identifying information, customer concern, related service history, cause, and correction.   Priority Rating 1

A.2.   Identify and interpret suspension and steering concern; determine necessary action.   Priority Rating 1

A.3.   Research applicable vehicle and service information, such as suspension and steering system operation, vehicle service history, service precautions, and technical service bulletins.   Priority Rating 1

A.4.   Locate and interpret vehicle and major component identification numbers (VIN, vehicle certification labels, calibration decals).   Priority Rating 1

B. Steering Systems Diagnosis and Repair

B.1.   Disable and enable supplemental restraint system (SRS).   Priority Rating 1

B.2.   Remove and replace steering wheel; center/time supplemental restraint system (SRS) coil (clock spring).   Priority Rating 1

B.3.   Diagnose steering column noises, looseness, and binding concerns (including tilt mechanisms); determine necessary action.   Priority Rating 2

B.4.   Diagnose power steering gear (non-rack-and-pinion) binding, uneven turning effort, looseness, hard steering, noises, and fluid leakage concerns; determine necessary action.   Priority Rating 3

B.5.   Diagnose power steering gear (rack-and-pinion) binding, uneven turning effort, looseness, hard steering, and fluid leakage concerns; determine necessary action.   Priority Rating 3

B.6.   Inspect steering shaft Universal joint(s), flexible coupling(s), collapsible column, lock cylinder mechanism, and steering wheel; perform necessary action.   Priority Rating 2

B.7.   Adjust manual or power non-rack-and-pinion worm bearing preload and sector lash.   Priority Rating 3

B.8.   Remove and replace manual or power rack-and-pinion steering gear; inspect mounting bushings and brackets.   Priority Rating 1

B.9.   Inspect and replace manual or power rack-and-pinion steering gear inner tie-rod ends (sockets) and bellows boots.   Priority Rating 1

B.10.   Determine proper power steering fluid type; inspect fluid level and condition.   Priority Rating 1

B.11.   Flush, fill, and bleed power steering system.   Priority Rating 2

B.12.   Diagnose power steering fluid leakage; determine necessary action.   Priority Rating 2

B.13.   Remove, inspect, replace, and adjust power steering pump belt.   Priority Rating 1

B.14.   Remove and reinstall power steering pump.   Priority Rating 3

B.15.   Remove and reinstall power steering pump pulley; check pulley and belt alignment.   Priority Rating 3

B.16.   Inspect and replace power steering hoses and fittings.   Priority Rating 2

B.17.   Inspect and replace pitman arm, relay (centerlink/intermediate) rod, idler arm and mountings, and steering linkage damper.   Priority Rating 2

B.18.   Inspect, replace, and adjust tie-rod ends (sockets), tie-rod sleeves, and clamps.   Priority Rating 1

B.19.   Diagnose and adjust components of electronically controlled
steering systems using a scan tool; determine necessary action.   Priority Rating 3

B.20.   Inspect and test non-hydraulic electric-power assist steering.   Priority Rating 3

B.21.   Identify hybrid vehicle power steering system electrical circuits,
service and safety precautions.   Priority Rating 3

C. Suspension Systems Diagnosis and Repair

1.   Front Suspension

C.1.1.   Diagnose short and long arm suspension system noises,
body sway, and uneven riding height concerns; determine
necessary action.   Priority Rating 1

C.1.2.   Diagnose strut suspension system noises, body sway, and
uneven riding height concerns; determine necessary action.   Priority Rating 1

C.1.3.   Remove, inspect, and install upper and lower control arms,
bushings, shafts, and rebound bumpers.   Priority Rating 3

C.1.4.   Remove, inspect, install, and adjust strut rods
(compression/tension) and bushings.   Priority Rating 2

C.1.5.   Remove, inspect, and install upper and lower ball joints.   Priority Rating 1

C.1.6.   Remove, inspect, and install steering knuckle assemblies.   Priority Rating 2

C.1.7.   Remove, inspect, and install short and long arm suspension
system coil springs and spring insulators.   Priority Rating 3

C.1.8.   Remove, inspect, install, and adjust suspension system torsion
bars; inspect mounts.   Priority Rating 3

C.1.9.   Remove, inspect, and install stabilizer bar bushings, brackets,
and links.   Priority Rating 2

C.1.10.   Remove, inspect, and install strut cartridge or assembly,
strut coil spring, insulators (silencers), and upper
strut bearing mount.   Priority Rating 1

C.1.11.   Lubricate suspension and steering systems.   Priority Rating 2

2.   Rear Suspension

C.2.1.   Remove, inspect, and install coil springs and spring insulators.   Priority Rating 2

C.2.2.   Remove, inspect, and install transverse links, control arms,
bushings, and mounts.   Priority Rating 2

C.2.3.   Remove, inspect, and install leaf springs, leaf spring insulators
(silencers), shackles, brackets, bushings, and mounts.   Priority Rating 3

C.2.4.   Remove, inspect, and install strut cartridge or assembly,
strut coil spring, and insulators (silencers).   Priority Rating 2

3.   Miscellaneous Service

C.3.1.   Inspect, remove, and replace shock absorbers.   Priority Rating 1

C.3.2.   Remove, inspect, and service or replace front and rear wheel
bearings.   Priority Rating 1

C.3.3.   Test and diagnose components of electronically controlled
suspension systems using a scan tool; determine necessary
action.   Priority Rating 3

D. Wheel Alignment Diagnosis, Adjustment, and Repair

D.1.   Diagnose vehicle wander, drift, pull, hard steering, bump steer,
memory steer, torque steer, and steering return concerns;
determine necessary action.   Priority Rating 1

D.2.   Perform prealignment inspection; perform necessary action.   Priority Rating 1

D.3.   Measure vehicle riding height; determine necessary action.   Priority Rating 1

D.4.   Check and adjust front and rear wheel camber; perform
necessary action.   Priority Rating 1

D.5.   Check and adjust caster; perform necessary action.   Priority Rating 1

D.6.	Check and adjust front wheel toe and center steering wheel.	Priority Rating 1
D.7.	Check toe-out-on-turns (turning radius); determine necessary action.	Priority Rating 2
D.8.	Check SAI (steering axis inclination) and included angle; determine necessary action.	Priority Rating 2
D.9.	Check and adjust rear wheel toe.	Priority Rating 1
D.10.	Check rear wheel thrust angle; determine necessary action.	Priority Rating 1
D.11.	Check for front wheel setback; determine necessary action.	Priority Rating 2
D.12.	Check front cradle (subframe) alignment; determine necessary action.	Priority Rating 3

E. Wheel and Tire Diagnosis and Repair

E.1.	Diagnose tire wear patterns; determine necessary action.	Priority Rating 1
E.2.	Inspect tires; check and adjust air pressure.	Priority Rating 1
E.3.	Diagnose wheel/tire vibration, shimmy, and noise; determine necessary action.	Priority Rating 2
E.4.	Rotate tires according to manufacturer's recommendations.	Priority Rating 1
E.5.	Measure wheel, tire, axle, and hub runout; determine necessary action.	Priority Rating 2
E.6.	Diagnose tire pull (lead) problem; determine necessary action.	Priority Rating 2
E.7.	Balance wheel and tire assembly (static and dynamic).	Priority Rating 1
E.8.	Dismount, inspect, and remount tire on wheel.	Priority Rating 2
E.9.	Dismount, inspect, and remount tire on wheel equipped with tire pressure sensor.	Priority Rating 3
E.10.	Reinstall wheel; torque lug nuts.	Priority Rating 1
E.11.	Inspect and repair tire and wheel assembly for air loss; perform necessary action.	Priority Rating 1
E.12.	Repair tire using internal patch.	Priority Rating 1
E.13.	Inspect, diagnose, and calibrate tire pressure monitoring system.	Priority Rating 1

DEFINITION OF TERMS USED IN THE TASK LIST

To clarify the intent of these tasks, NATEF has defined some of the terms used in the task list. To get a good understanding of what the task includes, refer to this glossary while reading the task list.

adjust	To bring components to specified operational settings.
align	To bring to precise alignment or relative position of components.
assemble (reassemble)	To fit together the components of a device.
balance	To establish correct linear, rotational, or weight relationship.
bleed	To remove air from a closed system.
check	To verify condition by performing an operational or comparative examination.
clean	To rid components of extraneous matter for the purpose of reconditioning, repairing, measuring, and reassembling.
determine	To establish the procedure to be used to effect the necessary repair.
determine necessary action	Indicates that the diagnostic routine or routines is the primary emphasis of a task. The student is required to perform the diagnostic steps and communicate the diagnostic outcomes and corrective actions required, addressing the concern or problem. The training program determines the communication method (worksheet, test, verbal communication, or other means deemed appropriate) and whether the corrective procedures for these tasks are actually performed.
diagnose	To locate the root cause or nature of a problem by using the specified procedure.

disassemble	To separate a component's parts in preparation for cleaning, inspection, or service.
fill (refill)	To bring fluid level to specified point or volume.
find	To locate a particular problem, such as shorts, grounds or opens in an electrical circuit.
flush	To use fluid to clean an internal system.
high voltage	Voltages of 50 volts or higher.
identify	To establish the identity of a vehicle or component before service; to determine the nature or degree of a problem.
inspect	(See *check*)
install (reinstall)	To place a component in its proper position in a system.
listen	To use audible clues in the diagnostic process; to hear the customer's description of a problem.
locate	Determine or establish a specific spot or area.
lubricate	To employ the correct procedures and materials in performing the prescribed service.
measure	To compare existing dimensions to specified dimensions by the use of calibrated instruments and gauges.
mount	To attach or place a tool or component in proper position.
on-board diagnostics (OBD)	A diagnostic system contained in the Powertrain Control Module (PCM), which monitors computer inputs and outputs for failures. OBD II is an industry-standard, second generation OBD system that monitors emissions control systems for degradation as well as failures.
perform	To accomplish a procedure in accordance with established methods and standards.
perform necessary action	Indicates that the student is to perform the diagnostic routine(s) and perform the corrective action item. Where various scenarious (conditions or situations) are presented in a single task, at least one of the scenarios must be accomplished.
pressure test	To use air or fluid pressure to determine the integrity, condition, or operation of a component or system.
proirity ratings	Indicates the minimum percentage of tasks, by area, a program must include in its curriculum in order to be certified in that area.
reassemble	(See *assemble*)
refill	(See *fill*)
remove	To disconnect and separate a component from a system.
repair	To restore a malfunctioning component or system to operating condition.
replace	To exchange an unserviceable component with a new or rebuilt component; to reinstall a component.
reset	(See *set*)
service	To perform a specified procedure when called for in the owner's manual or service manual.
set	To adjust a variable component to a given, usually initial, specification.
test	To verify a condition through the use of meters, gauges, or instruments.
torque	To tighten a fastener to a specified degree of tightness (in a given order or pattern if multiple fasteners are involved on a single component).

SUSPENSION AND STEERING SYSTEMS TOOLS AND EQUIPMENT

Many different tools and testing and measuring equipment are used to service suspension and steering systems. NATEF has identified many of these and has said that a suspension and steering technician must know what they are and how and when to use them. The tools and equipment listed by NATEF are covered in the following discussion. Also included are the tools and equipment you will use while completing the job sheets. Although you will be using common hand tools, they are not part of this discussion. You should already know what they are and how to use and care for them.

Stethoscope

Some sounds can be heard easily without using a listening device, but others are impossible to hear unless amplified. A stethoscope is very helpful in locating the cause of a noise by amplifying the sound waves. It can also help you distinguish between normal and abnormal noise. The procedure for using a stethoscope is simple. Use the metal prod to trace the sound until it reaches its maximum intensity. Once the precise location has been discovered, the sound can be better evaluated. A sounding stick, which is nothing more than a long, hollow tube, works on the same principle, although a stethoscope gives much clearer results.

The best results, however, are obtained with an electronic listening device. With this tool you can tune into the noise. Doing this allows you to eliminate all other noises that might distract or mislead you.

Belt Tension Gauge

A belt tension gauge is used to measure drive belt tension. The belt tension gauge is slipped over the belt, and the gauge indicates the amount of belt tension.

Tire Tread Depth Gauge

A tire tread depth gauge measures tire tread depth. This measurement should be taken at three or four locations around the tire's circumference to obtain an average tread depth. This gauge is used to determine the remaining life of a tire as well as for comparing the wear of one tire to the other tires. It is also used when making tire warranty adjustments.

Machinist's Rule

A machinist's rule is very much like an ordinary ruler. Each edge of this measuring tool is divided into increments based on a different scale. A typical machinist's rule based on the United States Customary System (USCS) of measurement may have scales based on 1/8-, 1/16-, 1/32-, and 1/64-inch intervals. Of course, metric machinist rules are also available. Metric rules are usually divided into 0.5-mm and 1-mm increments.

Some machinist's rules are based on decimal intervals. These are typically divided into 1/10-, 1/50-, and 1/1,000-inch (0.1, 0.05, and 0.001) increments. Decimal machinist's rules are very helpful when measuring dimensions that are specified in decimals, because you don't need to convert fractions to decimals.

Dial Indicator

The dial indicator is calibrated in 0.001-inch (one-thousandth-inch) increments. Metric dial indicators are also available. Both types are used to measure movement. Common uses of the dial indicator include measuring tire runout and ball joint movement. Dial indicators have many different attaching devices to connect the indicator to the component to be measured.

To use a dial indicator, position the indicator rod against the object to be measured. Then, push the indicator toward the work until the indicator needle travels far enough around the gauge face to permit movement to be read in either direction. Zero the indicator needle on the gauge. Move the object in the direction required while observing the needle of the gauge. Always be sure the range of the dial indicator is sufficient to allow the amount of movement required by the measuring procedure. For example, never use a 1-inch indicator on a component that will move 2 inches.

Power Steering Pressure Gauge

A power steering pressure gauge (Figure 2) is used to test the power steering pump pressure. Since the power steering pump delivers extremely high pressure during this test, the recommended procedure in the vehicle manufacturer's service manual must be followed. However, the typical

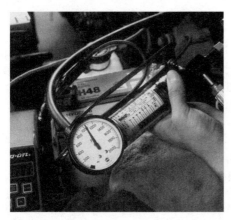

Figure 2 A power steering pressure tester.

procedure for using a pressure gauge on power steering systems is given here as an example.

To check the pressure of the pump, a pressure gauge with a shut-off valve is needed. With the engine off, disconnect the pressure hose at the pump. Install the pressure gauge between the pump and the steering gear. Use any adapters that may be necessary to make good connections with the vehicle's system. Open the shut-off valve and bleed the system as described in the service manual.

Start the engine and run it for approximately two minutes or until the engine reaches normal operating temperature. Then stop the engine and add fluid to the power steering pump if necessary. Now restart the engine and allow it to idle. Observe the pressure reading. The readings should be about 30 to 80 psi (200 to 550 kilo-Pascal [kPa]). If the pressure is lower than what is specified, the pump may be faulty. If the pressure is greater than specifications, the problem may be restricted hoses.

Now close the shut-off valve, observe the pressure reading, and reopen the valve. Do not keep the valve closed for more than 5 seconds. With the valve closed, the pressure should have increased to 600 to 1300 psi (4100 to 8950 kPa). Check the pressure reading on the gauge: if the pressure is too high, a faulty pressure relief valve is suggested. If the pressure is too low, the pump may be bad.

Scan Tools

The introduction of computer-controlled systems brought with it the need for tools capable of troubleshooting electronic control systems. A variety of computer scan tools are available today that do just that. A scan tool is a microprocessor designed to communicate with the vehicle's computer. Con-

nected to the computer through diagnostic connectors, a scan tool can access trouble codes, run tests to check system operations, and monitor the activity of the system. Trouble codes and test results are displayed on a light-emitting diode (LED) screen or printed out on the scanner printer.

Scan tools retrieve fault codes from a computer's memory and digitally display these codes on the tool. A scan tool may also perform many other diagnostic functions, depending on the year and make of the vehicle. Most aftermarket scan tools have removable modules that are updated each year. These modules are designed to test the computer systems on various makes of vehicles. For example, some scan testers have a 3-in-1 module that tests the computer systems on Chrysler, Ford, and General Motors vehicles. A 10-in-1 module is also available to diagnose computer systems on vehicles imported by 10 different manufacturers. These modules plug into the scan tool.

Scan tools are capable of testing many onboard computer systems, such as transmission controls, engine computers, antilock brake computers, air bag computers, and suspension computers, depending on the year and make of the vehicle and the type of scan tester. In many cases, the technician must select the computer system to be tested with the scanner after it has been connected to the vehicle.

The scan tool is connected to specific diagnostic connectors on various vehicles. Most manufacturers have one diagnostic connector. This connects the data wire from each onboard computer to a specific terminal in this connector. Other vehicle manufacturers have several different diagnostic connectors on each vehicle, each of which may be linked to one or more onboard computers. A set of connectors is supplied with the scanner to allow tester connection to various diagnostic connectors on different vehicles.

The scanner must be programmed for the model year, make of vehicle, and type of engine. With some scan tools, this selection is made by pressing the appropriate buttons on the tester, as directed by the digital tester display. On other scan testers, the appropriate memory card must be installed in the tester for the vehicle being tested. Some scan testers have a built-in printer to print test results, whereas other scan testers may be connected to an external printer.

As automotive computer systems become more complex, the diagnostic capabilities of scan testers continue to expand. Many scan testers are now able to store or "freeze" data in the tester during a road test and play it back when the vehicle is returned to the shop.

Some scan testers now display diagnostic information based on the fault code in the computer memory. The tester may index service bulletins published by the manufacturer of the scan tester after the vehicle information is entered in the tester. Other scan testers display sensor specifications for the vehicle being tested.

Trouble codes are set off by the vehicle's computer only when a voltage signal is entirely out of its normal range. The codes help technicians identify the cause of the problem. If a signal is within its normal range but is still not correct, the vehicle's computer does not display a trouble code; nonetheless, a problem still exists. To help identify this type of problem, most manufacturers recommend that the signals to and from the computer be carefully looked at. This is done with a scan tool or breakout box. A breakout box allows the technician to check voltage and resistance readings between specific points in the computer's wiring harness.

With On Board Diagnostic II (OBD-II), the diagnostic connectors are located in the same place on all vehicles. In addition, any scan tools designed for OBD-II work on all OBD-II systems; therefore, the need for designated scan tools or cartridges is eliminated. The OBD-II scan tool can run diagnostic tests on all systems and has "freeze frame" (storage) capabilities.

Floor Jack

A floor jack is a portable unit mounted on wheels. The lifting pad on the jack is placed under the chassis of the vehicle, and the jack handle is operated with a pumping action. This forces fluid into a hydraulic cylinder in the jack, and the cylinder extends to force the jack lift pad upward and lift the vehicle. Always be sure that the lift pad is positioned securely under one of the car manufacturer's recommended lifting points. To release the hydraulic pressure and lower the vehicle, the handle or release lever must be turned slowly.

The maximum lifting capacity of the floor jack is usually written on the jack decal. Never lift a vehicle that exceeds the jack lifting capacity. This action may cause the jack to break or collapse, resulting in vehicle damage or personal injury.

Lift

A lift is used to raise a vehicle so the technician can work under the vehicle. The lift arms must be placed under the car manufacturer's recommended lifting points prior to raising a vehicle. Twin posts are used on some lifts, whereas other lifts have a single post. Some lifts have an electric motor, which drives a hydraulic pump to create fluid pressure and force the lift upward. Other lifts use air pressure from the shop air supply to force the lift upward. If shop air pressure is used for this purpose, the air pressure is applied to fluid in the lift cylinder. A control lever or switch is placed near the lift. The control lever supplies shop air pressure to the lift cylinder, and the switch turns on the lift pump motor. Always be sure that the safety lock is engaged after the lift is raised. When the safety lock is released, a release lever is operated slowly to lower the vehicle.

Tire Changer

Tire changers are used to demount and mount tires (Figure 3). There are a wide variety of tire changers available, and each one has somewhat different operating procedures. Always follow the procedure in the equipment operator's manual and the directions provided by your instructor.

Wheel Balancer, Electronic Type

The most commonly used wheel balancer requires that the tire/wheel assembly be taken off and mounted on the balancer's spindle (Figure 4). A switch on the console sets the machine for either static or dynamic balancing. When the wheel balancing assembly is mounted for static balancing, it rotates until the heavy spot falls to the bottom. Weights are added to balance the assembly.

Several electronic dynamic/static balancer units permit balancing while the wheel and tire are on the car. Often a strobe light flashes at the heavy point of the tire and wheel assembly. On some machines, two separate tests must be done: one set of weights is placed to correct for static imbalance, and others are placed to correct for dynamic imbalance. Sometimes proper positioning of the static balance weights also corrects dynamic imbalance.

In the dynamic balance mode, the wheel assembly is rotated at high speed. Observing the balance scale, the operator reads out the amount

Figure 3 A tire changer is used to demount and mount tires.

Figure 4 An electronic wire balancer.

of weight that has to be added and the location where the weights should be placed.

Wheel Weight Pliers

Wheel weight pliers are actually combination tools designed to install and remove clip-on lead wheel weights. The jaws of the pliers are designed to hook into a hole in the weight's bracket. The pliers are then moved toward the outside of the wheel and the weight is pried off. On one side of the pliers is a plastic hammer head used to tap the weights onto the rim.

Bench Grinder

Bench grinders usually have a grinding wheel and a wire wheel brush driven by an electric motor. The grinding wheel may be replaced with a grinding disc containing several layers of synthetic material. A buffing wheel may be used in place of the wire wheel brush. The grinding wheel may be used for various grinding jobs and deburring. A buffing wheel is most commonly used for polishing.

Bench grinders must be securely bolted to the workbench. When grinding small components on a grinding wheel, wire brush wheel, or buffing wheel, always hold these components with a pair of vise grips to avoid injury to fingers and hands.

Hydraulic Press

When two components have a tight precision fit between them, a hydraulic press is used to either separate these components or press them together. The hydraulic press rests on the shop floor, and an adjustable steel beam bed is retained to the lower press frame with heavy steel pins. A hydraulic cylinder and ram are mounted on the top part of the press with the ram facing down, toward the press bed. The component being pressed is placed on the press bed with appropriate steel supports. A hand-operated hydraulic pump is mounted on the side of the press. When the handle is pumped, hydraulic fluid is forced into the cylinder, and the ram is extended against the component on the press bed to complete the pressing operation. A pressure gauge on the press indicates the pressure applied from the hand pump to the cylinder. The press frame is designed for a certain maximum pressure, which must not be exceeded during hand pump operation.

Handheld Grease Gun

A hand-operated grease gun (Figure 5) forces grease into a grease fitting. Often these are preferred because the technician controls the pressure of the grease. However, many shops use low air pressure to activate a pneumatic grease gun. The suspension and steering system may have several grease or zerk fittings.

Torque-Indicating Wrench

Torque is the twisting force used to turn a fastener against the friction between the threads and between the head of the fastener and the surface of the component. The fact that practically every vehicle and engine manufacturer publishes a list of torque recommendations is ample proof of the importance of using proper amounts of torque when tightening nuts or bolts. The amount of torque applied to a fastener is measured with a *torque-indicating* or *torque* wrench.

Three basic types of torque-indicating wrenches are available: a beam torque wrench has a beam that points to the torque reading, a click-type torque wrench has the desired torque reading set on the handle (when the torque reaches that level, the wrench clicks), and a dial torque wrench has a dial that indicates the torque exerted on the wrench. Some designs of the dial type

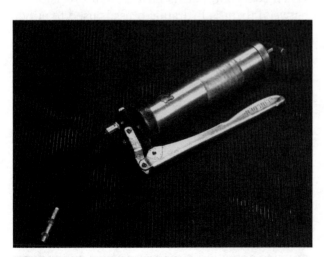

Figure 5 A handheld grease gun.

torque wrench have a light or buzzer that turns on when the desired torque is reached. All three types have pounds-per-inch and pounds-per-foot increments.

Gear and Bearing Pullers

Many tools are designed for a specific purpose. An example of a special tool is a gear and bearing puller. Many gears and bearings have a slight interference fit (press fit) when they are installed on a shaft or in a housing (a press fit is an interference fit). For example, the inside diameter of a bore is 0.001 inch smaller than the outside diameter of a shaft, so when the shaft is fitted into the bore it must be pressed in to overcome the 0.001-inch interference fit. This press fit prevents the parts from moving on each other. These gears and bearings must be removed carefully to prevent damage to the gears, bearings, and shafts. Prying or hammering can break or bind the parts. A puller with the proper jaws and adapters should be used to remove gears and bearings. With the proper puller, the force required to remove a gear or bearing can be applied with a slight and steady motion.

Bushing and Seal Pullers and Drivers

Another commonly used group of special tools are the various designs of bushing and seal drivers and pullers. Pullers are either a threaded or slide hammer type of tool. Always make sure you use the correct tool for the job because bushings and seals are easily damaged by the wrong tool or procedure. Car manufacturers and specialty tool companies work together closely to design and manufacture special tools required to repair cars. Most of these special tools are listed in the appropriate service manuals.

Seal drivers are designed to fit squarely against the seal case and inside the seal lip. A soft hammer is used to tap the seal driver and drive the seal straight into the housing. Some tool manufacturers market a seal driver kit with drivers to fit many common seals.

Tie-Rod End and Ball Joint Puller

Some car manufacturers recommend a tie rod end (Figure 6) and ball joint puller to remove tie rod

Figure 6 A tie rod end separating tool.

Figure 7 A pitman arm puller.

ends and pull ball joint studs from the steering knuckle. A tie rod end remover is a safer and better tool for separating ball joints than a pickle fork.

Ball joint removal and pressing tools are designed to remove and replace pressed-in ball joints on front suspension systems. Often these tools are used in conjunction with a hydraulic press. The size of the removal and pressing tool must match the size of the ball joint.

Some ball joints are riveted to the control arm and the rivets are drilled out for removal.

Pitman Arm Puller

A pitman arm puller (Figure 7) is a heavy-duty puller designed to remove the pitman arm from the pitman shaft. Never strike a puller with a hammer when it is installed and tightened.

Axle Pullers

Axle pullers are used to pull rear axles in rear-wheel-drive vehicles. Most rear axle pullers are slide hammer-type.

Control Arm Bushing Tools

A variety of control arm bushing tools are available to remove and replace control arm bushings. Old bushings are pressed out of the control arm. A C-clamp tool can be used to remove the bushing. The C-clamp is installed over the bushing. An adapter is selected to fit on the bushing and push

the bushing through the control arm. Turning the handle on the C-clamp pushes the bushing out of the control arm.

New bushings can be installed by driving or pressing them in place. Adapters are available for the C-clamp tool to install the new bushings. After the correct adapters are selected, position the bushing and tool on the control arm. Turning the C-clamp handle pushes the bushing into the control arm.

Front Bearing Hub Tool

Front bearing hub tools are designed to remove and install front wheel bearings on front-wheel-drive cars. These bearing hub tools are usually designed for a specific make of vehicle and the correct tools must be used for each application. Failure to do so may result in damage to the steering knuckle or hub, not to mention that the use of the wrong tool will waste quite a bit of your time.

Coil Spring Compressor Tool

Many types of coil spring compressor tools are available to the automotive service industry (Figure 8). These tools are designed to compress the coil spring and hold it in the compressed position while removing the strut from the coil spring

Figure 8 A coil spring compressor in place around a coil spring.

Figure 9 A special spring compressor for MacPherson struts.

(Figure 9), removing the spring from an SLA suspension, or performing other suspension work. Various types of spring compressor tools are required on different types of front suspension systems.

One type of spring compressor uses a threaded compression rod that fits through two plates, an upper and lower ball nut, a thrust washer, and a forcing nut. The two plates are positioned at either end of the spring. The compression rod fits through the plates with a ball nut at either end. The upper ball nut is pinned to the rod. The thrust washer and forcing nut are threaded onto the end of the rod. Turning the forcing nut draws the two plates together and compresses the spring.

A compressed coil spring contains a tremendous amount of energy. Never disconnect any suspension component that will suddenly release this tension; this action may result in serious personal injury and vehicle or property damage.

Shock Absorber Tools

Often shock absorbers can be removed with regular hand tools, but there are times when special tools may be necessary. The shocks are under the vehicle and so are subject to dirt and moisture,

which may make it difficult to loosen the mounting nut from the stud of the shock. Wrenches are available to hold the stud while loosening the nut. There are also tools for pneumatic chisels to help work off the nut.

Power Steering Pump Pulley Special Tool Set

When a power steering pump pulley must be replaced, it should never be hammered off or on—doing this causes internal damage to the pump. Normally the pulley can be removed with a gear puller, although special pullers are available. To install a pulley, a special tool is used to press the pulley on without using a press or needing to drive the pulley into place.

Steering Column Special Tool Set

A wheel puller is used to remove the steering wheel from its shaft. Mount the puller over the wheel's hub after the horn button and air bag have been removed. Make sure you follow the recommendations exactly for air bag module removal. Screw the bolts into the threaded bores in the steering wheel. Then tighten the puller's center bolt against the steering wheel shaft until the steering wheel is free.

Special tools are also required to service the lock mechanism and ignition switch.

Tie Rod Sleeve-Adjusting Tool

A tie rod sleeve-adjusting tool is required to rotate the tie-rod sleeves and perform some front wheel adjustments. Never use anything except a tie rod adjusting tool to adjust the tie-rod sleeves. Tools such as pipe wrenches damage the sleeves.

Turning Radius Gauge

Turning radius gauge turntables are placed under the front wheels during a wheel alignment. The top plate in the turning radius gauge rotates on the bottom plate to allow the front wheels to be turned during a wheel alignment. A degree scale and a pointer on the gauge indicate the number of degrees the front wheels are turned. If the car has four-wheel steering, the turning radius gauges are also placed under the rear wheels during a wheel alignment.

Plumb Bob

A plumb bob is a metal weight with a tapered end suspended on a string. Plumbers use a plumb bob to locate pipe openings directly below each other at the top and bottom of partitions. Some vehicle manufacturers recommend checking vehicle frame measurements with a plumb bob.

Tram Gauge

A tram gauge is a long, straight, graduated bar with an adjustable pointer at each end. It is used for performing frame and body measurements.

Magnetic Wheel Alignment Gauge

The magnetic wheel alignment gauge is capable of measuring some of the front suspension alignment angles. Each magnetic wheel alignment gauge contains a strong magnet that holds the gauge securely on the front wheel hubs. Magnetic wheel alignment gauge mounting surfaces must be clean with no metal burrs.

Rim Clamps

When the wheel hub is inaccessible to the magnetic alignment gauge, an adjustable rim clamp may be attached to each front wheel. The magnetic gauges may be attached to the rim clamp. Rim clamps are also used on computer wheel aligners.

Brake Pedal Depressor

A brake pedal depressor must be installed between the front seat and the brake pedal to keep the vehicle from moving while checking front wheel alignment angles.

Steering Wheel Locking Tool

A steering wheel locking tool is required to lock the steering wheel while performing some front suspension service and during certain steps of the alignment process. Its purpose is to prevent the steering wheel and the wheels from turning. When adjusted properly, the steering wheel lock also helps keep the steering wheel centered during alignment.

Toe Gauge

A toe gauge is a long, straight, graduated bar that may be used to measure front wheel toe.

Track Gauge

Some track gauges use a fiber-optic alignment system to measure front wheel toe and to determine if the rear wheels are tracking directly behind the front wheels. The front and rear fiber-optic gauges may be connected to the wheel hubs or to rim clamps attached to the wheel rims. A remote light source in the main control is sent through fiber-optic cables to the wheel gauges. A strong light beam between the front and rear wheel units informs the technician if the rear wheel tracking is correct.

Computer Wheel Aligner

Many automotive shops are equipped with a computer wheel aligner. These wheel aligners perform checks on all front and rear wheel alignment angles quickly and accurately. A typical computerized system displays information on a cathode ray tube (CRT) screen to guide the technician step by step through the alignment process.

After vehicle information is keyed into the machine and the wheel units are installed (Figure 10), the machine must be compensated for wheel runout. When compensation is complete, alignment measurements are instantly displayed, along with the specifications for that vehicle. In addition to the normal alignment specifications, the CRT may display asymmetric tolerances, different left- and right-side specifications, and cross specifications (difference allowed between left and right side). Graphics and text on the screen show the technician where and how to make adjustments. As the adjustments are made on the vehicle, the technician can observe the center block slide toward the target. When the block aligns with the target, the adjustment is within half the specified tolerance.

Service Manuals

Perhaps the most important tools you will use are service manuals. There is no way a technician can remember all the procedures and specifications to repair all vehicles. Thus a good technician relies

on service manuals and other information sources. Good information plus knowledge allows a technician to fix a problem with the least frustration and at the lowest expense to the customer.

To obtain the correct specifications and other information, you must first identify the vehicle you are working on. The best source for positive identification is the VIN.

The primary source of repair and specification information for any car, van, or truck is the manufacturer. The manufacturer publishes service manuals each year, for every vehicle built. Because of the enormous amount of information, some manufacturers publish more than one manual per year per car model. They are typically divided into sections based on the major systems of the vehicle. Manufacturers' manuals cover all repairs, adjustments, specifications, detailed diagnostic procedures, and special tools required.

Since many technical changes occur on specific vehicles each year, manufacturers' service manuals need to be constantly updated. Updates are published as service bulletins (often referred to as Technical Service Bulletins or TSBs) that show the changes in specifications and repair procedures dur-

Figure 10 The wheel unit for an electronic wheel alignment machine.

ing the model year. These changes do not appear in the service manual until the next year. The car manufacturer provides these bulletins to dealers and repair facilities on a regular basis.

Service manuals are also published by independent companies rather than the manufacturers. However, they pay for and get most of their information from the car makers. They contain component information, diagnostic steps, repair procedures, and specifications for several car makes in one book. Information is usually condensed and is more general in nature than the manufacturer's manuals. The condensed format allows for more coverage in less space and, therefore, is not always specific. They may also contain several years of models as well as several car makes in one book.

Many of the larger parts manufacturers have excellent guides on the various parts they manufacture or supply. They also provide updated service bulletins on their products. Other sources for up-to-date technical information are trade magazines and trade associations.

The same information that is available in service manuals is now commonly found electroni-cally on compact disks (CD-ROMs), digital video disks (DVDs), and the Internet. A single compact disk can hold a quarter million pages of text, eliminating the need for a huge library to contain all of the printed manuals. Using electronics to find information is also easier and quicker. The disks are normally updated quarterly and not only contain the most recent service bulletins but also engineering and field service fixes. DVDs can hold more information than CDs; therefore, fewer disks are needed with systems that use DVDs. The CDs and DVDs are inserted into a computer. All a technician needs to do is enter vehicle information and then move to the appropriate part or system. The appropriate information will then appear on the computer's screen. Online data can be updated instantly and requires no space for physical storage. These systems are easy to use and the information is quickly accessed and displayed. The computer's keyword, mouse, and/or light pen are used to make selections from the screen's menu. Once the information is retrieved, a technician can read it off the screen or print it out and take it to the service bay.

CROSS-REFERENCE GUIDE

NATEF Task	Job Sheet
A.1	1
A.2	2
A.3	3
A.4	3
B.1	4
B.2	4
B.3	5
B.4	6
B.5	7
B.6	8
B.7	9
B.8	10
B.9	11 & 12
B.10	13
B.11	14
B.12	15
B.13	16
B.14	17
B.15	16
B.16	17
B.17	18
B.18	19
B.19	20
B.20	21
B.21	22
C.1.1	23
C.1.2	24
C.1.3	25
C.1.4	26
C.1.5	27
C.1.6	28
C.1.7	29
C.1.8	30
C.1.9	31
C.1.10	32
C.1.11	33
C.2.1	34

JOB SHEETS

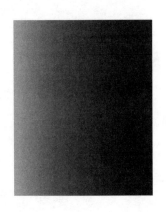

SUSPENSION AND STEERING JOB SHEET 1

Filling Out a Work Order

Name _____ Station _____ Date _____

NATEF Correlation

This Job Sheet addresses the following NATEF task:

A.1. Complete work order to include customer information, vehicle identifying information, customer concern, related service history, cause, and correction.

Objective

Upon completion of this job sheet, you will be able to prepare a service work order based on customer input, vehicle information, and service history.

Tools and Materials

An assigned vehicle or the vehicle of your choice

Service work order or computer-based shop management package

Parts and labor guide

Work Order Source: Describe the system used to complete the work order. If a paper repair order is being used, describe the source.

_____

_____

PROCEDURE

1. Prepare the shop management software for entering a new work order or obtain a blank paper work order.   Task Completed ☐

2. Enter customer information, including name, address, and phone numbers onto the work order.   Task Completed ☐

3. Locate and record the vehicle's VIN.   Task Completed ☐

4. Enter the necessary vehicle information, including year, make, model, engine type and size, transmission type, license number, and odometer reading.   Task Completed ☐

5. Does the VIN verify that the information about the vehicle is correct?

_____

6. Normally, you would interview the customer to identify his or her concerns. However to complete this job sheet, assume the only concern is a bouncy ride and the apparent cause is bad rear shock absorbers. This concern should be added to the work order.   Task Completed ☐

7. The history of service to the vehicle can often help diagnose problems as well as indicate possible premature part failure. Gathering this information from the customer can provide some of this information. For this job sheet assume the vehicle has not had a similar problem and was not recently involved in a collision. Service history is further obtained by searching files for previous service. Often this search is done by customer name, VIN, and license number. Check the files for any related service work. Task Completed ☐

8. Search for technical service bulletins on this vehicle that may relate to the customer's concern. Task Completed ☐

9. Based on the customer's concern, service history, TSBs, and your knowledge, what is the likely cause of this concern?

   _____

   _____

10. Enter this information onto the work order. Task Completed ☐

11. Prepare to make a repair cost estimate for the customer. Identify all parts that may need to be replaced to correct the concern. List these here.

    _____

    _____

12. Describe the task(s) that will be necessary to replace the part.

    _____

    _____

13. Using the parts and labor guide, locate the cost of the parts that will be replaced and enter the cost of each item onto the work order at the appropriate place for creating an estimate. Task Completed ☐

14. Now, locate the flat rate time for work required to correct the concern. List each task and with its flat rate time.

    _____

    _____

15. Multiply the time for each task by the shop's hourly rate and enter the cost of each item onto the work order at the appropriate place for creating an estimate. Task Completed ☐

16. Many shops have a standard amount they charge each customer for shop supplies and waste disposal. For this job sheet, use an amount of ten dollars for shop supplies. Task Completed ☐

17. Add the total costs and insert the sum as the subtotal of the estimate. Task Completed ☐

18. Taxes must be included in the estimate. What is the sales tax rate and does it apply to both parts and labor, or just one of these?

    _____

    _____

19. Enter the appropriate amount of taxes to the estimate, than add this to the subtotal. The end result is the estimate to give the customer.

Task Completed ☐

20. By law, how accurate must your estimate be?

_____

_____

21. Generally speaking, the work order is complete and is ready for the customer's signature. However, some businesses require additional information; make sure you enter that information to the work order. On the work order there is a legal statement that defines what the customer is agreeing to. Briefly describe the contents of that statement.

_____

_____

Problems Encountered

_____

_____

_____

Instructor's Comments

_____

_____

_____

SUSPENSION AND STEERING JOB SHEET 2

Identifying Problems and Concerns

Name _____ Station _____ Date _____

NATEF Correlation

This Job Sheet addresses the following NATEF task:

A.2. Identify and interpret suspension and steering concern; determine necessary action.

Objective

Upon completion of this job sheet, you will be able to define suspension and steering system problems or concerns, prior to diagnosing or testing the systems.

Tools and Materials

None required

Protective Clothing

Goggles or safety glasses with side shields

Describe the vehicle being worked on:

Year _____ Make _____ Model _____

VIN _____ Engine type and size _____

PROCEDURE

1. Start the engine and describe how the engine seems to be running.

   _____

   _____

2. Take the vehicle for a safe road test and pay strict attention to how the steering system operates in all conditions (straight-ahead, slow-partial turn, fast-partial turn, slow-full turn, and fast-full turn), in both directions. Describe your results here.

   _____

   _____

   _____

   _____

3. Were there any unusual noises when the steering wheel was turned? If so, describe them and when they occurred.

_____

_____

_____

_____

_____

_____

4. Were there any unusual noises from the vehicle as it went over bumps or went into curves? If so, describe them and when they occurred.

_____

_____

5. Describe how the steering wheel felt during the road test.

_____

_____

6. Did the vehicle tend to pull to one side when the brakes were applied? Explain.

_____

_____

7. Visually inspect all tires. What did you find and what does that indicate?

_____

_____

_____

_____

8. Check the power steering fluid and describe the level and condition of the fluid.

_____

_____

9. Describe the type of front and rear suspension this vehicle has.

_____

_____

_____

_____

10. Describe the type of steering system this vehicle has.

   _____

   _____

11. Based on the above, what are your suspicions and conclusions about the suspension and steering systems?

   _____

   _____

   _____

   _____

Problems Encountered

_____

_____

_____

Instructor's Comments

_____

_____

_____

SUSPENSION AND STEERING JOB SHEET 3

Gathering Vehicle Information

Name _____ Station _____ Date _____

NATEF Correlation

This Job Sheet addresses the following NATEF tasks:

A.3. Research applicable vehicle and service information, such as suspension and steering system operation, vehicle service history, service precautions, and technical service bulletins.

A.4. Locate and interpret vehicle and major component identification numbers (VIN, vehicle certification labels, calibration labels).

Objective

Upon completion of this job sheet, you will be able to gather service information about a vehicle and its suspension and steering system.

Tools and Materials

Appropriate service manuals

Computer

Protective Clothing

Goggles or safety glasses with side shields

Describe the vehicle being worked on:

Year _____ Make _____ Model _____

VIN _____

PROCEDURE

1. Using the service manual or other information source, describe what each letter and number in the VIN for this vehicle represents.

   _____

   _____

2. Locate the Vehicle Emissions Control Information (VECI) label and describe where you found it.

   _____

   _____

3. Summarize the information you found on the VECI label.

   _____

   _____

4. Using a service manual or electronic database, locate the information about the vehicle's suspension and steering system. List the major components of the systems and describe how they are controlled, if they are.

_____

_____

_____

_____

_____

5. Using a service manual or electronic database, locate and record all service precautions regarding the suspension and steering system noted by the manufacturer.

_____

_____

_____

_____

6. Using the information that is available, locate and record the vehicle's service history.

_____

_____

_____

_____

7. Using the information sources that are available, summarize all Technical Service Bulletins for this vehicle that relate to the suspension and steering systems.

_____

_____

_____

_____

Problems Encountered

_____

_____

Instructor's Comments

_____

_____

SUSPENSION AND STEERING JOB SHEET 4

Working Safely Around Air Bags

Name _____ Station _____ Date _____

NATEF Correlation

This Job Sheet addresses the following NATEF tasks:

B.1. Disable and enable supplemental restraint system (SRS).

B.2. Remove and replace steering wheel; center/time supplemental restraint system (SRS) coil (clock spring).

Objective

Upon completion of this job sheet, you will be able to disable and enable supplemental restraint systems (SRS). You will also be able to properly remove and replace the steering wheel and center or time the SRS coil.

Tools and Materials

A vehicle with air bags

Service manual for the above vehicle

Digital multimeter (DMM)

Component locator for the above vehicle

Protective Clothing

Goggles or safety glasses with side shields

Describe the vehicle being worked on:

Year _____ Make _____ Model _____

VIN _____ Engine type and size _____

List all of the restraint systems found on this vehicle:

_____

_____

_____

PROCEDURE

1. Locate the information about the air bag system in the service manual. How are the critical parts of the system identified in the vehicle?

_____

_____

_____

_____

2. List the main components of the air bag system and describe their location.

_____

_____

_____

_____

3. Here are some very important guidelines to follow when working with and around air bag systems. These are listed below with some key words left out. Read through these and fill in the blanks with the correct words.

   a. Wear _____ _____ when servicing an air bag system and when handling an air bag module.

   b. Wait at least _____ minutes after disconnecting the battery before beginning any service. The reserve _____ module is capable of storing enough energy to deploy the air bag for up to _____ minutes after battery voltage is lost.

   c. Always handle all _____ and other components with extreme care. Never strike or jar a sensor, especially when the battery is connected. Doing so can cause deployment of the air bag.

   d. Never carry an air bag module by its _____ or _____, and, when carrying it, always face the trimmed side of the module _____ from your body. When placing a module on a bench, always face the trimmed side of the module _____.

   e. Deployed air bags may have a powdery residue on them. _____ _____ is produced by the deployment reaction and is converted to _____ _____ when it comes in contact with the moisture in the atmosphere. Although it is unlikely that harmful chemicals will still be on the bag, it is wise to wear _____ _____ and _____ when handling a deployed air bag. Immediately wash your hands after handling a deployed air bag.

   f. A live air bag must be _____ before it is disposed. A deployed air bag should be disposed of in a manner consistent with the _____ and manufacturer's procedures.

   g. Never use a battery- or AC-powered _____, _____, or any other type of test equipment in the system unless the manufacturer specifically says to. Never probe with a _____ _____ for voltage.

Problems Encountered

_____

_____

_____

Instructor's Comments

_____

_____

SUSPENSION AND STEERING JOB SHEET 5

Checking the Steering Column

Name _____ Station _____ Date _____

NATEF Correlation

This Job Sheet addresses the following NATEF task:

B.3. Diagnose steering column noises, looseness, and binding concerns (including tilt mechanisms); determine necessary action.

Objective

Upon completion of this job sheet, you will be able to diagnose steering column noises, looseness, and binding concerns, including tilt mechanisms.

Tools and Materials

Service manual

Protective Clothing

Goggles or safety glasses with side shields

Describe the vehicle being worked on:

Year _____ Make _____ Model _____

VIN _____ Engine type and size _____

PROCEDURE

1. Sitting behind the steering wheel with the ignition key out, attempt to rotate the steering wheel. Can you turn it? _____

2. Put the key in and turn it to the on position. Can you turn the steering wheel? _____ Is it difficult to move? _____

3. Turn the key off and remove it. Was it difficult to move? _____

4. Summarize the condition of the steering column lock mechanism.

   _____

   _____

5. If the steering column is equipped with a telescoping or tilt feature, move the steering wheel through the complete available range. Does it move easily through the range? _____ Does it lock into all positions?

   _____

6. Summarize the condition of the tilt or telescoping steering wheel feature.

   _____

   _____

7. With the key in the on position, attempt to move the steering wheel toward you and away from you. Describe how much movement there is.

_____

_____

8. Now move the steering wheel side to side without turning the wheels. This is the amount of play in the steering column assembly. Summarize your findings.

_____

_____

9. Prepare to take the vehicle for a road test.   Task Completed ☐

10. Drive the vehicle at normal speeds and pay attention to the feel and sounds of the steering wheel and column. Summarize your findings.

_____

_____

11. What services do you recommend for the steering column?

_____

_____

_____

_____

Problems Encountered

_____

_____

_____

Instructor's Comments

_____

_____

_____

SUSPENSION AND STEERING JOB SHEET 6

Diagnosing a Power Recirculating Ball Steering Gear

Name _____ Station _____ Date _____

NATEF Correlation

This Job Sheet addresses the following NATEF task:

B.4. Diagnose power steering gear (non-rack-and-pinion) binding, uneven turning effort, looseness, hard steering, noises, and fluid leakage concerns; determine necessary action.

Objective

Upon completion of this job sheet, you will be able to diagnose a power recirculating ball steering gear for noise, binding, uneven turning effort, looseness, fluid leakage, and hard steering.

Tools and Materials

Basic hand tools

Protective Clothing

Goggles or safety glasses with side shields

Describe the vehicle being worked on:

Year _____ Make _____ Model _____

VIN _____ Engine type and size _____

PROCEDURE

1. Take the vehicle on a road test and pay attention to the noises and action of the steering system. Describe the results of the test.

   _____

   _____

   _____

   _____

2. If the steering gear was noisy, check these items:

   a. Loose pitman shaft lash adjustment—may cause a rattling noise when the steering wheel is turned.

   b. Cut or worn dampener O-ring on the valve spool—when this defect is present, a squawking noise is heard during a turn.

   c. Loose steering gear mounting bolts.

   d. Loose or worn flexible coupling or steering shaft U-joints.

   e. A hissing noise from the power steering gear is normal if the steering wheel is at the end of its travel, or when the steering wheel is rotated with the vehicle standing still.

   Task Completed ☐

3. Summarize the above checks.

   _____

   _____

   _____

   _____

4. If the steering wheel jerked or surged when the steering wheel was turned with the engine running, check the power steering pump belt condition and tension. Describe the results of the test.

   _____

   _____

5. If excessive kickback was felt on the steering wheel, check the poppet valve in the steering gear. Describe the results of the test.

   _____

   _____

6. If the steering felt loose, check these defects:

   a. Air in the power steering system. To remove the air, fill the power steering pump reservoir and rotate the steering wheel fully in each direction several times.

   b. Loose pitman lash adjustment.

   c. Loose worm shaft thrust bearing preload adjustment.

   d. Worn flexible coupling or universal joint.

   e. Loose steering gear mounting bolts.

   f. Worn steering gears.   Task Completed ☐

7. Summarize the above checks.

   _____

   _____

   _____

   _____

8. If the steering felt hard while parking, check for the following:

   a. Loose or worn power steering pump belt.

   b. Low oil level in the power steering pump.

   c. Excessively tight steering gear adjustments.

   d. Defective power steering pump with low-pressure output.

   e. Restricted power steering hoses.

   f. Defects in the steering gear such as: pressure loss in the cylinder because   Task Completed ☐
      of scored cylinder, worn piston ring, or damaged backup O-ring, excessively loose spool in the valve body, or defective or improperly installed gear check poppet valve.

9. Summarize the above checks.

_____

_____

_____

_____

Problems Encountered

Instructor's Comments

SUSPENSION AND STEERING JOB SHEET 7

Diagnosing a Power Rack-and-Pinion Steering Gear

Name _____ Station _____ Date _____

NATEF Correlation

This Job Sheet addresses the following NATEF task:

B.5. Diagnose power steering gear (rack-and-pinion) binding, uneven turning effort, looseness, hard steering, and fluid leakage concerns; determine necessary action.

Objective

Upon completion of this job sheet, you will be able to diagnose and inspect a rack and pinion type power steering system and determine what is necessary to correct any problems.

Tools and Materials

Service manual

Protective Clothing

Goggles or safety glasses with side shields

Describe the vehicle being worked on:

Year _____ Make _____ Model _____

VIN _____ Engine type and size _____

PROCEDURE

Diagnosis

1. Take the vehicle on a road test and pay attention to the noises and action of the steering system. Describe the results of the test.

   _____

   _____

   _____

   _____

2. If the steering gear was noisy, check these items:

   • Loose steering gear mounting bolts.

   • Loose or worn flexible coupling or steering shaft U-joints.

   • A hissing noise from the power steering gear is normal if the steering wheel is at the end of its travel, or when the steering wheel is rotated with the vehicle standing still.

- A groaning or howling can indicate a restriction in the hydraulic system.

- A chirping or squealing normally means the drive belt is loose or slipping.

- Improperly routed hoses, loose mounting bolts on the power steering pump or the steering gear can cause a rattle.   Task Completed ☐

3. Summary of the above checks.

   _____

   _____

   _____

   _____

4. If the steering wheel jerked or surged when the steering wheel was turned with the engine running, check the power steering pump belt condition and tension. Describe the results of the test.

   _____

   _____

5. If excessive feedback was felt through the steering wheel, check for loose or damaged steering gear mounts and tie rod ends.

   _____

   _____

6. If the steering felt loose, check these defects:

   - Air in the power steering system. To remove the air, fill the power steering pump reservoir and rotate the steering wheel fully in each direction several times.

   - Worn flexible coupling or universal joint.

   - Loose steering gear mounting bolts.

   - Loose, worn, or damaged tie rod ends.   Task Completed ☐

7. Summary of the above checks.

   _____

   _____

   _____

   _____

8. If the steering felt hard while parking, check for the following:

   - Loose or worn power steering pump belt.

   - Low oil level in the power steering pump.

- Defective power steering pump with low-pressure output.
- Restricted power steering hoses.
- Damaged or faulty steering column bearings.
- Seized steering column U-joints.
- Internal problems in the power rack assembly.
- Faulty suspension components.
- Inadequately inflated tires.

Task Completed ☐

9. Summary of the above checks.

_____

_____

_____

_____

Visual Inspection

1. Inspect the tires. Check for correct pressure, construction, size, wear, and damage, and for defects that include ply separations, sidewall knots, concentricity problems, and force problems. Describe what you found.

_____

_____

2. Check the drive belt for the power steering pump and adjust belt tension if necessary. Describe what you found.

_____

_____

3. Check the fluid level and condition. Describe what you found.

_____

_____

4. With the ignition OFF, wipe off the outside of the power steering pump, pressure hose, return hose, and the steering gear.

5. Start the engine and turn the steering wheel several times from stop to stop. Check for leaks. Describe what you found.

_____

_____

6. With the front wheels straight ahead and the engine stopped, measure the maximum steering wheel freeplay and compare your findings to specifications. Describe what you found and explain what this indicates.

_____

_____

7. With the vehicle sitting on the shop floor and the front wheels straight ahead, have an assistant turn the steering wheel about 1/4 turn in both directions. Watch for looseness in the steering shaft's flexible coupling or universal joints. Describe what you found and explain what this indicates.

_____

_____

8. Grasp the pinion gear shaft at the flexible steering coupling and try to move it in and out of the gear. Describe what you found and explain what this indicates.

_____

_____

9. Carefully inspect the rack housing. Have an assistant turn the steering wheel about 1/2 turn in both directions and watch for movement of the housing in its mounting bushings. Describe what you found and explain what this indicates.

_____

_____

10. Inspect the bellows boots for cracks, splits, leaks, and proper clamp installation. Describe what you found.

_____

_____

11. Check the inner tie-rod socket assemblies inside the bellows. Squeeze the boot until the inner socket can be felt. Then, push and pull on the tire. Describe what you found and explain what this indicates.

_____

_____

12. While an assistant turns the steering wheel 1/4 turn in each direction, watch for looseness in the outer tie-rod ends. Describe what you found and explain what this indicates.

_____

_____

13. While conducting a dry park check, grab each tie rod end and feel for any roughness that would indicate internal rusting. Check the outer tie-rod end seals for cracks and proper installation of the nuts and cotter pins. Describe what you found and explain what this indicates.

_____

_____

14. Inspect the tie rods for a bent condition. Describe what you found and explain what this indicates.

_____

_____

15. Summarize the condition of the steering gear.

_____

_____

_____

_____

Problems Encountered

_____

_____

_____

Instructor's Comments

_____

_____

_____

SUSPENSION AND STEERING JOB SHEET 8

Inspect, Remove, and Replace Steering Column

Name _____ Station _____ Date _____

NATEF Correlation

This Job Sheet addresses the following NATEF task:

B.6. Inspect steering shaft universal-joint(s), flexible coupling(s), collapsible column, lock
cylinder mechanism, and steering wheel; perform necessary action.

Objective

Upon completion of this job sheet, you will be able to inspect, remove, and replace the components of
a steering column.

Tools and Materials

Seat cover

Torque wrench

Protective Clothing

Goggles or safety glasses with side shields

Describe the vehicle being worked on:

Year _____ Make _____ Model _____

VIN _____ Engine type and size _____

PROCEDURE

Inspection

1. Does the vehicle show signs of repaired or unrepaired collision damage?
   Look at thee service history and carefully check the body for evidence.
   Describe your findings.

   _____

   _____

2. Does the steering column stay steady during a test drive? Does it seem
   loose during turns or when driving straight? Describe your findings.

   _____

   _____

3. Does the column appear to be the source of rattles while driving straight on irregular road surfaces?

_____

_____

4. Inspect the components of the energy-absorbing steering column. Describe your findings.

_____

_____

5. With the vehicle's weight resting on the front suspension, watch the flexible coupling or universal joint as an assistant turns the steering wheel 1/2 turn in each direction. If the vehicle has power steering, the engine should be running with the gear selector in Park. What did you observe?

_____

_____

6. Based on these checks, what services do you recommend?

_____

_____

Remove and Replace

1. Disconnect the battery's negative cable. If the vehicle is equipped with an air bag, wait for the time period specified by the vehicle manufacturer before progressing through this job sheet. What is the specified waiting period to prevent accidental air bag deployment? Why so long?

_____

_____

2. Install a seat cover on the front seat.   Task Completed ☐

3. Place the front wheels in the straight-ahead position, and remove the ignition key from the switch to lock the steering column.   Task Completed ☐

4. Remove the cover under the steering column and remove the lower finish panel if necessary.   Task Completed ☐

5. Disconnect all wiring connectors from the steering column.   Task Completed ☐

6. If the vehicle has a column-mounted gearshift lever, disconnect the gearshift linkage at the lower end of the steering column.   Task Completed ☐

7. Inspect the lock cylinder mechanism and any mechanical linkage to the lock. Then disconnect that linkage. Describe the condition of the lock and linkage.

_____

_____

8. Remove the retaining bolt or bolts in the lower universal joint or flexible coupling.   Task Completed ☐

9. Inspect the universal joint and/or the flexible couplings of the column. Describe their condition.

    _____

    _____

10. Inspect the collapsible column unit. Describe its condition.

    _____

    _____

11. Remove the steering-column-to-instrument-panel mounting bolts.   Task Completed ☐

12. Carefully remove the steering column from the vehicle. Be careful not to damage upholstery or paint.   Task Completed ☐

13. List all of the parts that need to be replaced during assembly.

    _____

    _____

14. Install the steering column under the instrument panel, and insert the steering shaft into the lower universal joint.   Task Completed ☐

15. Install the steering-column-to-instrument-panel mounting bolts. Be sure the steering column is properly positioned, and tighten these bolts to the specified torque. What is the torque specification?

    _____

    _____

16. Install the retaining bolt or bolts in the lower universal joint or flexible coupling, and tighten the bolt(s) to the specified torque. What is the torque specification?

    _____

    _____

17. Connect the lock mechanism linkage.   Task Completed ☐

18. Connect the gearshift linkage, if the vehicle has column mounted gearshift.   Task Completed ☐

19. Connect all the wiring harness connectors to the steering column connectors.   Task Completed ☐

20. Install the steering column cover and the lower finish panel.   Task Completed ☐

21. Reconnect the negative battery cable.   Task Completed ☐

22. Road test the vehicle and check for proper steering column operation. Give a summary of its operation.

    _____

    _____

Problems Encountered

_____

_____

_____

Instructor's Comments

_____

_____

_____

SUSPENSION AND STEERING JOB SHEET 9

Adjusting a Recirculating Ball Steering Gear

Name _____ Station _____ Date _____

NATEF Correlation

This Job Sheet addresses the following NATEF task:

B.7. Adjust manual or power non–rack-and-pinion worm bearing preload and sector lash.

Objective

Upon completion of this job sheet, you will be able to adjust manual or power recirculating ball steering gear bearing preload and sector lash.

Tools and Materials

Torque wrench, in-lbs

Protective Clothing

Goggles or safety glasses with side shields

Describe the vehicle being worked on:

Year _____ Make _____ Model _____

VIN _____ Engine type and size _____

Make of steering gear _____

PROCEDURE

Typical Manual Steering Systems

1. Describe the results of loose steering gear adjustments on steering quality.

   _____

   _____

2. Describe the results of tight steering gear adjustments.

   _____

   _____

3. Loosen the worm shaft adjuster plug lock nut with a brass punch and a hammer, and tighten the adjuster plug until all the worm shaft endplay is removed, then loosen the plug one-quarter turn.   Task Completed ☐

4. Turn the worm shaft fully to the right with a socket and an inch-pound torque wrench, then turn the worm shaft one-half turn toward the center position.   Task Completed ☐

5. Tighten the adjuster plug until the specified bearing preload is indicated on the torque wrench as the worm shaft is rotated. The specification on some steering gears is 5 to 8 in-lbs (0.56 to 0.896 Nm). Always use the vehicle manufacturer's specified preload. What are the exact specifications?

_____

6. Tighten the adjuster plug lock nut to 85 ft-lbs (114 Nm).   Task Completed ☐

7. Turn the pitman backlash adjuster screw outward until it stops, then turn it inward one turn.   Task Completed ☐

8. Rotate the worm shaft fully from one stop to the other stop, and carefully count the number of shaft rotations. The total number of worm shaft turns was?

_____

9. Turn the worm shaft back exactly one-half the total number of turns from one of the stops. How many turns did you move the worm shaft from the fully right or left position to the center position?

_____

10. With the steering gear positioned as it was in step 9, connect an inch-pound torque wrench and socket to the worm shaft, and note the steering gear turning torque while rotating the worm shaft 45 degrees in each direction. What were your findings?

_____

_____

11. Turn the pitman backlash adjuster screw until the torque wrench reading is 6 to 10 in-lbs (0.44 to 1.12 Nm) more than the worm shaft bearing preload torque in step 5.   Task Completed ☐

12. Tighten the pitman backlash adjuster screw lock nut to the specified torque. What are the exact specifications?

_____

Typical Power Steering Systems

1. Remove the worm shaft thrust bearing adjuster plug lock nut with a hammer and brass punch. What are the exact specifications?

_____

2. Turn this adjuster plug inward, or clockwise, until it bottoms and tighten the plug to 20 ft-lbs (27 Nm).   Task Completed ☐

3. Place an index mark on the steering gear housing next to one of the holes in the adjuster plug.   Task Completed ☐

4. Measure 0.50 in (13 mm) counterclockwise from the index mark, and place a second index mark at this position.   Task Completed ☐

5. Rotate the adjuster plug counterclockwise until the hole in the adjuster plug is aligned with the second index mark placed on the housing.

Task Completed ☐

6. Install and tighten the adjuster plug lock nut to the specified torque. What are the exact specifications?

_____

7. State any problems you encountered while completing this job sheet.

_____

_____

_____

Problems Encountered

_____

_____

_____

Instructor's Comments

_____

_____

_____

SUSPENSION AND STEERING JOB SHEET 10

Servicing a Rack-and-Pinion Steering Gear

Name _____ Station _____ Date _____

NATEF Correlation

This Job Sheet relates to the following NATEF task:

B.8. Remove and replace manual or power rack-and-pinion steering gear; inspect mounting bushings and brackets.

Objective

Upon completion of this job sheet, you will be able to disassemble, inspect, adjust, repair, and reassemble a rack-and-pinion steering gear.

Tools and Materials

Soft-jaw vise   Snap ring pliers

Basic hand tools   Anhydrous calcium grease

Pull scale   Lithium-based grease

Protective Clothing

Goggles or safety glasses with side shields

Describe the vehicle being worked on:

Year _____ Make _____ Model _____

VIN _____ Engine type and size _____

PROCEDURE

Inspection of Components

1. If the pinion bearing is loose on the pinion shaft, replace the pinion and bearing assembly. Describe its condition.

   _____

   _____

2. Check the pinion shaft for worn or chipped teeth. Describe its condition.

   _____

   _____

3. Inspect the pilot bearing contact area on the pinion shaft. Wear, pitting, or scoring in this area indicates that a new pinion shaft is required. Describe its condition.

   _____

   _____

4. Check the pinion bearing for roughness, looseness, and noisy operation. Describe its condition.

_____

_____

5. Check the pilot bearing in the steering gear housing. If this bearing is worn or scored, replace the pilot bearing and the pinion shaft and bearing assembly. Describe its condition.

_____

_____

6. Check the rack bushing for excessive wear. Describe its condition.

_____

_____

7. Check the mounting bushings for looseness and excessive wear. If the bushings are loose or worn, replace them. Always replace the bushings in pairs. If the bushings are in satisfactory condition, do not disturb them. Describe their condition.

_____

_____

To Assemble the Gear

1. Install the rack in the housing with the teeth parallel to the pinion shaft.   Task Completed ☐

2. Move the rack until the specified distance is obtained between the end of the rack and the housing. If this specification is not available, position the rack ends at the distance from the housing recorded during the disassembly procedure.

3. Use an anhydrous calcium grease to lubricate the pilot bearing and the pilot bearing contact area on the pinion shaft.   Task Completed ☐

4. Install the pinion shaft with the punch marks aligned on the shaft and housing.   Task Completed ☐

5. Install the pinion shaft retaining ring with the beveled edge facing upward.   Task Completed ☐

6. Place a generous coating of anhydrous calcium grease on top of the pinion shaft bearing.   Task Completed ☐

7. Use the proper seal driver to install the pinion shaft seal.   Task Completed ☐

8. Apply a coating of lithium-based grease to the rack bearing and install this bearing.   Task Completed ☐

9. Apply lithium-based grease to the ends of the preload spring and the adjuster plug threads, and install the preload spring and adjuster plug.   Task Completed ☐

10. Turn the adjuster plug until it bottoms, then rotate the plug 45 to 60 degrees counterclockwise.   Task Completed ☐

11. Rotate the pinion shaft with a socket and a torque wrench and observe the turning torque on the torque wrench. If necessary, rotate the adjuster plug to obtain the specified turning torque. Tighten the adjuster plug lock nut to the specified torque. What is the specification?

_____

12. Coat both ends of the rack with lithium-based grease and fill the rack teeth with the same lubricant. Rotate the pinion shaft fully in each direction several times. Apply more grease to the rack teeth after each complete pinion shaft rotation.

Task Completed ☐

13. Rotate the inner tie-rod ends onto the rack until they bottom.

Task Completed ☐

14. Install the inner tie-rod pins or stake these ends as required. Always use a wooden block to support the opposite side of the rack and inner tie-rod end while staking. If jam nuts are located on the inner tie-rod ends, be sure they are tightened to the specified torque. What is the specification?

_____

15. Place a large bellows boot clamp over each end of the gear housing. Make sure the boots are seated in the housing and the tie-rod undercuts. Install and tighten the large inner boot clamps.

Task Completed ☐

16. Install the outer bellows boot clamps on the tie rods, but do not install these clamps on the boots until the steering gear is installed and the toe is adjusted.

Task Completed ☐

17. Install the jam nuts and the outer tie-rod ends. Align the marks placed on these components during the disassembly procedure. Leave the jam nuts loose until the steering gear is installed and the toe is adjusted.

Task Completed ☐

18. Install the steering gear in the vehicle and check the front wheel toe. Tighten the outer bellows boot clamps and the outer tie-rod end jam nuts.

Task Completed ☐

Problems Encountered

_____

_____

_____

Instructor's Comments

_____

_____

_____

SUSPENSION AND STEERING JOB SHEET 11

Inspect a Rack-and-Pinion Steering Gear

Name _____ Station _____ Date _____

NATEF Correlation

This Job Sheet addresses the following NATEF task:

B.9. Inspect and replace manual or power rack-and-pinion steering gear inner tie rod ends (sockets) and bellows boots.

Objective

Upon completion of this job sheet, you will be able to demonstrate the ability to remove and replace a manual or power rack-and-pinion steering gear and tie rods. You will also be able to inspect the mounting bushings and brackets.

Tools and Materials

Tape measure

Protective Clothing

Goggles or safety glasses with side shields

Describe the vehicle being worked on:

Year _____ Make _____ Model _____

VIN _____ Engine type and size _____

Describe general operating condition:

_____

PROCEDURE

1. With the front wheels straight ahead and the engine stopped, rock the steering wheel gently back and forth with light finger pressure and measure the maximum steering wheel freeplay with a tape measure.

   Specified steering wheel freeplay: _____

   Measured steering wheel freeplay: _____

2. What must be done in order to bring abnormal amounts of freeplay back within specifications?

   _____

   _____

   _____

3. With the vehicle sitting on the shop floor and the front wheels straight ahead, have an assistant turn the steering wheel about one-quarter turn in both directions. Watch for looseness in the flexible coupling or steering column U-joints and the steering shaft. Describe your findings:

_____

_____

_____

4. While an assistant turns the steering wheel about one-half turn in both directions, watch for movement of the steering gear housing in its mounting bushings. Describe your findings:

_____

_____

_____

5. Grasp the pinion shaft extending from the steering gear and attempt to move it vertically. Describe your findings:

_____

_____

_____

6. What needs to be done if there is movement?

_____

_____

_____

7. Road test the vehicle and check for excessive steering effort. Describe your findings:

_____

_____

_____

8. What could cause excessive steering effort?

_____

_____

_____

9. Visually inspect the bellows boots for cracks, splits, leaks, and proper clamp installation. Replace any boot that is damaged. Describe the condition of both the right boot and left boot:

_____

_____

10. Loosen the inner bellows boot clamps and move each boot toward the outer tie-rod end until the inner tie-rod end is visible. Push outward and inward on each front tire and watch for movement in the inner tie-rod end. An alternate method of checking the inner tie-rod end is to squeeze the bellows boot and grasp the inner tie-rod end socket. Movement in the inner tie-rod end is then felt as the front wheel is moved inward and outward. Hard plastic bellows boots may be found on some applications. With this type of bellows boot, remove the ignition key from the switch to lock the steering column and push inward and outward on the front tire while observing any lateral movement in the tie rod. Describe your findings:

_____

_____

_____

11. Grasp each outer tie-rod end and check for vertical movement. While an assistant turns the steering wheel one-quarter turn in each direction, watch for looseness in the outer tie-rod ends. Check the outer tie-rod end seals for cracks and proper installation of the nuts and cotter pins. Inspect the tie-rods for a bent condition. Describe what you found:

_____

_____

_____

Problems Encountered

_____

_____

_____

Instructor's Comments

_____

_____

_____

SUSPENSION AND STEERING JOB SHEET 12

Servicing Inner Tie-Rod Ends on a Rack-and-Pinion Steering Gear

Name _____ Station _____ Date _____

NATEF Correlation

This Job Sheet addresses the following NATEF task:

B.9. Inspect and replace manual or power rack-and-pinion steering gear inner tie-rod ends (sockets) and bellows boots.

Objective

Upon completion of this job sheet, you will be able to demonstrate the ability to inspect, remove, and replace an inner tie-rod end on a manual or power rack-and-pinion steering gear.

Tools and Materials

Vise   Punch set

Hand tools   Service manual

Protective Clothing

Goggles or safety glasses with side shields

Describe the vehicle being worked on:

Year _____ Make _____ Model _____

VIN _____ Engine type and size _____

PROCEDURE

1. With the rack removed from the vehicle and mounted securely in a soft-jawed vise, place an index mark on the outer tie-rod end, jam nut, and tie rod.   Task Completed ☐

2. Loosen the jam nut and remove the jam nut and outer tie-rod end.   Task Completed ☐

3. Remove the inner and outer boot clamps.   Task Completed ☐

4. Remove the boot from the tie rod.   Task Completed ☐

5. Hold the rack with the properly sized wrench and loosen the inner tie-rod end with the appropriate wrench.   Task Completed ☐

6. Remove the inner tie rod from the rack.   Task Completed ☐

7. Inspect the tie rod and describe its condition.

_____

_____

8. Make sure the shock damper ring is in place, then install the inner tie rod into the rack and tighten it to specifications while holding the rack in place with another wrench.

Task Completed ☐

9. Support the inner tie rod over the vise, and stake both sides of the inner tie-rod joint with a hammer and a punch.

Task Completed ☐

10. Use a feeler gauge to measure the clearance between the rack and the inner tie-rod joint housing stake. What is the specification for this clearance?

_____

11. Compare your measurement to the specifications and summarize your recommendations based on that comparison.

_____

_____

_____

_____

12. Install a new boot and boot clamps.

Task Completed ☐

13. Install the jam nut and outer tie-rod end. Make sure the index marks are aligned before tightening the jam nut.

Task Completed ☐

Problems Encountered

_____

_____

_____

Instructor's Comments

_____

_____

_____

SUSPENSION AND STEERING JOB SHEET 13

Inspect the Power Steering Fluid

Name _____ Station _____ Date _____

NATEF Correlation

This Job Sheet addresses the following NATEF task:

B.10. Determine proper power steering fluid type; inspect fluid level and condition.

Objective

Upon completion of this job sheet, you will be able to check the power steering fluid levels and condition.

Tools and Materials

White rag
Clean power steering fluid
Service manual

Protective Clothing

Goggles or safety glasses with side shields

Describe the vehicle being worked on:

Year _____ Make _____ Model _____

VIN _____ Engine type and size _____

PROCEDURE

1. Refer to the service manual and state the type of lubricant that should be used in this power steering system.

   _____

   _____

2. Locate the power steering fluid reservoir and describe its location.

   _____

   _____

3. With the engine idling at about 1000 rpm or less, turn the steering wheel slowly and completely in each direction several times. Why is this step important?

   _____

   _____

   _____

4. If the vehicle has a remote power steering fluid reservoir, check the fluid for signs of foaming. If the fluid is foamy, what is indicated?

_____

_____

_____

5. Check the level of the fluid in the remote reservoir. What did you find?

_____

_____

6. Shut off the engine and remove the dirt from around the cap and neck of the fluid reservoir.

Task Completed ☐

7. Remove the cap and check the fluid level. Is the level correct? _____

8. Check the condition of the fluid by smelling it and wiping a sample on a clean white rag. Based on these checks, what are your conclusions about the fluid and the power steering system?

_____

_____

_____

9. Fill the reservoir to appropriate level with the correct fluid.

Task Completed ☐

Problems Encountered

_____

_____

_____

Instructor's Comments

_____

_____

_____

SUSPENSION AND STEERING JOB SHEET 14

Flush a Power Steering System

Name _____ Station _____ Date _____

NATEF Correlation

This Job Sheet addresses the following NATEF task:

B.11. Flush, fill, and bleed power steering system.

Objective

Upon completion of this job sheet, you will be able to drain, flush, and refill a power steering system.

Tools and Materials

Floor jack and jack stands or hoist

Drain pan

Hand tools

Power steering fluid

Protective Clothing

Goggles or safety glasses with side shields

Describe the vehicle being worked on:

Year _____ Make _____ Model _____

VIN _____ Engine type and size _____

Describe general operating condition:

_____

PROCEDURE

1. Describe the steering problems and noises that occur when there is air in the power-steering system.

   _____

   _____

   _____

2. Describe the steering problems and noises that occur when power-steering fluid is contaminated.

   _____

   _____

   _____

3. Remove the return hose from the power-steering pump reservoir. Place a plug in the reservoir outlet and position the open return line hose end into a drain pan.   Task Completed ☐

4. With the engine idling, turn the steering wheel fully in each direction and stop the engine.   Task Completed ☐

5. Fill the reservoir to the hot full mark with the correct power-steering fluid. The recommended fluid type is: _____

6. Start the engine and run it at 1000 rpm while observing the return hose in the drain pan. When the fluid begins to discharge from the return hose, shut off the engine.   Task Completed ☐

7. Repeat steps 4, 5, and 6 until there is no air or dirty fluid discharging from the return hose. How many times did you need to repeat the process? _____

8. Remove the plug from the reservoir and reconnect the return hose.   Task Completed ☐

9. Fill the reservoir to the specified level.   Task Completed ☐

10. With the engine running at 1000 rpm, turn the steering wheel fully in each direction three or four times. Each time the steering wheel is turned fully in a direction, hold it there for 2 to 3 seconds before turning it in the opposite direction.   Task Completed ☐

11. Check for foaming of the fluid in the reservoir. When foaming is present, repeat steps 9 and 10 to remove the air trapped in the system.   Task Completed ☐

12. Check the fluid level and make sure it is at the hot full mark.   Task Completed ☐

Problems Encountered

_____

_____

_____

Instructor's Comments

_____

_____

_____

SUSPENSION AND STEERING JOB SHEET 15

Diagnose Power Steering Leakage Problems

Name _____ Station _____ Date _____

NATEF Correlation

This Job Sheet addresses the following NATEF task:

B.12. Diagnose power steering fluid leakage; determine necessary action.

Objective

Upon completion of this job sheet, you will be able to identify the cause of power rack-and-pinion steering fluid leakage problems.

Tools and Materials

Power steering fluid

Protective Clothing

Goggles or safety glasses with side shields

Describe the vehicle being worked on:

Year _____ Make _____ Model _____

VIN _____ Engine type and size _____

PROCEDURE

CAUTION: *If the engine has been running for a length of time, power steering gears, pumps, lines, and fluid may be very hot. Wear eye protection and protective gloves when servicing these components.*

1. Make sure the power steering reservoir is filled to the specified level with the proper power steering fluid.

   Task Completed ☐

2. Make sure the power steering fluid is at normal operating temperature. If necessary, rotate the steering wheel several times from lock-to-lock to bring the fluid to normal operating temperature.

   Task Completed ☐

3. Inspect the fluid reservoir for signs of leaks. If evidence of leakage is found, describe where the fluid appears to be coming from.

   _____

   _____

4. Inspect the power steering pump for signs of leakage. If evidence of leakage is found, describe where the fluid appears to be coming from.

   _____

   _____

5. Carefully check the power steering hoses and fittings at the pump for signs of leakage. Do this check while an assistant turns the wheels with the engine running. If evidence of leakage is found, describe where the fluid appears to be coming from.

   _____

   _____

6. Check the hoses themselves for leaks while an assistant turns the wheels with the engine running. If evidence of leakage is found, describe where the fluid appears to be coming from.

   _____

   _____

7. Inspect the cylinder end of the steering gear for oil leaks. If evidence of leakage is found, describe where the fluid appears to be coming from.

   _____

   _____

8. Inspect the steering gear for oil leaks at the pinion end of the housing with the engine running and the steering wheel turned so the rack is against the left internal stop. If evidence of leakage is found, describe where the fluid appears to be coming from.

   _____

   _____

9. Inspect the pinion end of the power steering gear housing for oil leaks with the steering wheel turned in either direction. If evidence of leakage is found, describe where the fluid appears to be coming from.

   _____

   _____

10. Inspect the pinion coupling area for oil leaks. If evidence of leakage is found, describe where the fluid appears to be coming from.

    _____

    _____

11. Inspect all the lines and fittings on the steering gear for oil leaks as an assistant turns the wheels with the engine running. If evidence of leakage is found, describe where the fluid appears to be coming from.

    _____

    _____

12. With the engine running, turn the steering wheel fully in each direction and check the steering effort. Describe how the turning effort feels and if it feels the same in both directions.

    _____

    _____

13. Summarize your conclusions about the power steering system.

_____

_____

_____

Problems Encountered

_____

_____

_____

Instructor's Comments

_____

_____

_____

SUSPENSION AND STEERING JOB SHEET 16

Power Steering Pump Belt and Pulley Service

Name _____ Station _____ Date _____

NATEF Correlation

This Job Sheet addresses the following NATEF tasks:

B.13. Remove, inspect, replace, and adjust power steering pump belt.

B.15. Remove and reinstall power steering pump pulley; check pulley and belt alignment.

Objective

Upon completion of this job sheet, you will be able to remove, inspect, and replace a power steering pump, mounts, seals, and gaskets, as well as its drive belt and pulley and check the alignment of the pulley and belt.

Tools and Materials

Machinist rule	Straightedge
Puller assortment	Belt tension gauge
Vise	Pry bar
Holding tool	

Protective Clothing

Goggles or safety glasses with side shields

Describe the vehicle being worked on:

Year _____ Make _____ Model _____

VIN _____ Engine type and size _____

PROCEDURE

1. Check the alignment of the power steering belt and pulley by positioning a long straightedge in parallel with the belt and pulley of any component driven by the belt. Look down and over the straightedge to see if all of the components are aligned. Record your findings.

   _____

   _____

2. Explain the service required to align power steering pump pulley with other related pulleys and give the reasons for your diagnosis.

   _____

   _____

3. Inspect the power steering belt for fraying, oil soaking, wear on friction surfaces, cracks, glazing, and splits. If the belt is damaged or defective, it must be replaced. Record your findings.

_____

_____

4. If the belt is good, check its tension. Press on the belt at the longest belt span with the engine stopped to measure the belt deflection, which should be 1/2 inch per foot of free span. At what length of belt span did you measure belt deflection? How much deflection did you measure?

_____

_____

5. Install a belt tension gauge over the belt in the center of the longest span to measure the belt tension. What is the specified belt tension? How much did you measure?

_____

_____

6. Start the engine and observe the movement of the pulley. If the pulley wobbles while it is rotating, the pulley is probably bent, and pulley replacement is necessary. Record your findings.

_____

_____

7. Turn off the engine and check for worn belt grooves in the pulley. This problem also dictates pulley replacement. Record your findings.

_____

_____

8. Check the pulley for cracks. If this condition is present, pulley replacement is essential. Record your findings.

_____

_____

9. Loosen the power steering pump bracket or tension adjusting bolt. Check the bracket and pump mounting bolts for wear. Record your findings.

_____

_____

10. List the required power steering pump and bracket service and explain the reasons for your diagnosis.

_____

_____

11. Remove the power steering belt.   Task Completed ☐

12. To remove a pulley that is pressed onto the pump shaft, use a special puller   Task Completed ☐
    to remove the pulley and a pulley installation tool to install it.

13. If the power steering pump pulley is retained with a nut, mount the pump
    in a vise. Tighten the vise on one of the pump mounting bolt surfaces. Do
    not tighten the vise with excessive force. Install the special holding tool to
    keep the pulley from turning, and loosen the pulley nut with a box end
    wrench. Remove the nut, pulley, and woodruff key. Inspect the pulley, shaft,
    and woodruff key for wear. Be sure the key slots in the shaft and pulley are
    not worn. Replace all worn components. Record your findings.

    _____

    _____

14. Install the pulley.   Task Completed ☐

15. Install the new power steering pump belt over all the pulleys.   Task Completed ☐

16. Pry against the pump ear and hub with a pry bar to tighten the belt. Some   Task Completed ☐
    pump brackets have a 1/2-inch square opening in which a breaker bar may
    be installed to move the pump and tighten the belt.

17. Hold the pump in the position described in step 13, and tighten the brack-   Task Completed ☐
    et or tension adjusting bolt.

18. Recheck the belt tension with the tension gauge. What is the specified
    belt tension? How much did you measure?

    _____

    _____

19. Tighten the tension adjusting bolt and the mounting bolts to the manu-
    facturer's specified torque. What are the specifications for the power
    steering pump mounting bolts and the power steering pump bracket or
    tension adjusting bolt?

    _____

    _____

Problems Encountered

_____

_____

_____

Instructor's Comments

_____

_____

_____

SUSPENSION AND STEERING JOB SHEET 17

Inspect and Replace Power Steering Hoses, Lines, and Pumps

Name _____ Station _____ Date _____

NATEF Correlation

This Job Sheet addresses the following NATEF tasks:

B.14. Remove and reinstall power steering pump.

B.16. Inspect and replace power steering hoses and fittings.

Objective

Upon completion of this job sheet, you will be able to inspect and replace power steering hoses, lines, and fittings.

Tools and Materials

Drain pan

Clean power-steering fluid

Protective Clothing

Goggles or safety glasses with side shields

Protective gloves

Describe the vehicle being worked on:

Year _____ Make _____ Model _____

VIN _____ Engine type and size _____

PROCEDURE

1. With the engine running and an assistant turning the steering wheel from side-to-side, check for leaks at the hoses and lines and their connections. Record the results of this check.

   _____

   _____

2. Check also for signs of hose ballooning while the steering wheel is being moved. Record the results of this check.

   _____

   _____

3. Turn off the engine and visually check the hoses for damage. Record the results of this check.

   _____

   _____

4. Check the lines for dents, sharp bends, cracks, and contact with other components. Record the results of this check.

_____

_____

CAUTION: *If the engine has been running, the power-steering hoses, components, and fluid may be extremely hot. Wear protective gloves and use caution to avoid burns.*

5. Summarize your findings and list the hoses or lines that need to be replaced.

_____

_____

_____

6. To replace a hose or line, turn the engine off and remove the return hose at the power-steering gear.   Task Completed ☐

7. Allow the fluid to drain from this hose into a drain pan.   Task Completed ☐

8. Loosen and remove all hose fittings from the pump and steering gear.   Task Completed ☐

9. Remove all hose-to-chassis clips.   Task Completed ☐

10. Remove the hoses from the chassis, and cap the pump and steering gear fittings.   Task Completed ☐

11. If O-rings are used on the hose ends, install new O-rings. If the system uses gaskets, pry out the old gasket from the fitting in the housing before the new lines and gaskets are installed. Lubricate O-rings with power-steering fluid.   Task Completed ☐

12. Reverse steps 6 through 10 to install the power-steering hoses.   Task Completed ☐

13. Tighten all fittings to the manufacturer's specified torque. What are those specifications?

_____

14. Make sure the hoses don't rub on other parts, and that all hose-to-chassis clips are in place.   Task Completed ☐

15. Fill the pump reservoir to the full mark with the manufacturer's recommended fluid. What type of fluid does this system use?

_____

16. Bleed air from the power-steering system. Check the fluid level in the reservoir and add fluid as required.   Task Completed ☐

Removing and Replacing the Pump

1. The procedure for removing the pump will vary from vehicle model to model. Briefly describe the procedure required to gain access to the pump for removal.

_____

_____

2. Drain power steering fluid.   Task Completed ☐

3. Remove the drive belt.   Task Completed ☐

4. Disconnect the oil reservoir to pump hose.   Task Completed ☐

5. Disconnect the connector from the power steering oil pressure switch.   Task Completed ☐

6. Disconnect the pressure feed tube assembly.   Task Completed ☐

7. Remove the pump assembly.   Task Completed ☐

8. To install the pump, place it at its proper location and tighten the mounting bolts to specifications. What are the torque specifications?

   _____

9. Connect the connector to the power steering oil pressure switch.   Task Completed ☐

10. Install the pressure feed tube assembly, with a new gasket, to the pump. Make sure the tube is properly placed in the pump. Then tighten the fitting to specifications. What are the specifications?

    _____

11. Connect the oil reservoir to pump hose.   Task Completed ☐

12. Install the drive belt and adjust to the correct tension. What is the specification for belt tension?

    _____

13. Reconnect all items that were removed or disconnected to gain access to the pump.   Task Completed ☐

14. Fill the reservoir with the correct fluid.   Task Completed ☐

15. Then, bleed the power steering system.   Task Completed ☐

16. Run the engine and inspect the system for leaks.   Task Completed ☐

17. State any difficulties you had while removing and replacing the pump.

    _____

    _____

Problems Encountered

_____

_____

_____

Instructor's Comments

_____

_____

_____

SUSPENSION AND STEERING JOB SHEET 18

Diagnose, Remove, and Replace Steering Linkage Components

Name _____ Station _____ Date _____

NATEF Correlation

This Job Sheet addresses the following NATEF task:

B.17. Inspect and replace pitman arm, relay (center link/intermediate) rod, idler arm and mountings, and steering linkage damper.

Objective

Upon completion of this job sheet, you will be able to inspect and replace a pitman arm, relay (center link/intermediate) rod, idler arm and mountings, and steering linkage damper. •

Tools and Materials

Pull scale   Hand tools

Dial indicator   Grease gun

Torque wrench

Protective Clothing

Goggles or safety glasses with side shields

Describe the vehicle being worked on:

Year _____ Make _____ Model _____

VIN _____ Engine type and size _____

Describe general operating condition:

_____

PROCEDURE

1. Attach the magnetic base of a dial indicator to the frame near the idler arm.   Task Completed ☐

2. Position the dial indicator's plunger against the upper side of the outer end of the idler arm. Preload the dial indicator and set the dial to zero.   Task Completed ☐

3. Use a pull scale to apply 25 pounds (11.334 kg) of force downward and then upward on the idler arm. Observe the total movement on the dial indicator. Compare this reading to the specifications for idler arm vertical movement.

   Specified vertical idler arm movement _____

   Measured vertical idler arm movement _____

4. Record your conclusions and recommendations from this test:

_____

_____

5. To remove the idler arm, begin by removing the idler arm to center link cotter pin and nut.

Task Completed ☐

6. Remove the center link from the idler arm.

Task Completed ☐

7. Remove the idler arm bracket mounting bolts, and remove the idler arm.

Task Completed ☐

8. If the idler arm has a steel bushing, thread the bracket into the idler arm bushing until the specified clearance is obtained between the center of the lower bracket bolt hole and the upper idler arm surface.

Task Completed ☐

The specified distance from the center of the lower bracket hole to the upper idler arm surface is:

_____

9. Install the idler arm bracket to frame bolts and tighten them to the specified torque. Be sure that lock washers are installed on the bolts.

Task Completed ☐

The specified idler arm retaining bolt torque is:

_____

10. Install the center link into the idler arm and tighten the mounting nut to the specified torque. Install the cotter pin into the nut after tightening.

Task Completed ☐

The specified center link to idler arm nut torque is:

_____

11. If the idler arm has a grease fitting, lubricate it as required.

Task Completed ☐

Problems Encountered

_____

_____

_____

Instructor's Comments

_____

_____

_____

SUSPENSION AND STEERING JOB SHEET 19

Remove and Replace an Outer Tie-Rod End on a Parallelogram Steering Linkage

Name _____ Station _____ Date _____

NATEF Correlation

This Job Sheet addresses the following NATEF task:

B.18. Inspect, replace, and adjust tie-rod ends (sockets), tie-rod sleeves, and clamps.

Objective

Upon completion of this job sheet, you will be able to remove, inspect, adjust, and replace tie-rod sleeves and clamps on a parallelogram-type steering linkage.

Tools and Materials

Hand tools

Tie rod puller

Drive on lift/hoist

Protective Clothing

Goggles or safety glasses with side shields

Describe the vehicle being worked on:

Year _____ Make _____ Model _____

VIN _____ Engine type and size _____

Describe general operating condition:

_____

PROCEDURE

1. Raise the vehicle on a lift. If the vehicle has rubber encapsulated tie-rod ends, make sure the wheels are pointed straight ahead.

   Task Completed ☐

2. Loosen the nut on the tie-rod sleeve bolt that retains the sleeve to the tie-rod ends.

   Task Completed ☐

3. Measure from the top center of the tie-rod end ball stud to the outer end of the tie-rod sleeve. Your measurement was: _____

4. Remove the cotter pin from the tie-rod end retaining nut.

   Task Completed ☐

5. Loosen the tie-rod end retaining nut.

   Task Completed ☐

6. With a tie-rod end puller, loosen the tie-rod end in the steering arm.

   Task Completed ☐

7. Remove the tie-rod end retaining nut. Then remove the tie-rod end from the steering arm.

   Task Completed ☐

8. Count the number of turns required to thread the tie-rod end from the tie-rod sleeve and record them here: _____

9. Install the new tie-rod end into the tie-rod sleeve, turning it the same number of turns counted in step #8.   Task Completed ☐

10. Push the tie-rod end ball stud into its bore in the steering arm. Only the threads on the tie-rod end ball stud should be visible above the steering arm surface.

    If this is not the case, what should you do?

    _____

    _____

11. With the tie-rod end ball stud pushed fully into the steering arm, measure the distance from the center of the ball stud to the outer end of the tie-rod end sleeve. Your measurement is: _____

12. If this measurement is *not* the same as the measurement recorded in step #3, remove the tie-rod end ball stud from the steering arm and rotate the tie-rod end until this distance equals the measurement taken in step #3.   Task Completed ☐

13. Install the tie-rod end ball stud retaining nut and tighten it to the specified torque. The specified torque is: _____

14. Install a new cotter pin through the tie-rod end ball stud and nut openings and bend the ends of the cotter pin separately around the nut.   Task Completed ☐

    WARNING: *Never loosen a tie-rod end ball stud to align the openings in the ball stud and nut to allow cotter pin installation. This can cause the nut to loosen during vehicle operation, which will result in excessive steering freeplay. Complete loss of steering control will occur if the nut comes off the tie-rod end ball stud.*

15. Tighten the nut on the outer tie-rod sleeve clamp nut bolt to the specified torque. The specified torque is: _____

    Make sure the slot in the tie-rod sleeve is positioned away from the opening of the clamp.

16. Measure and correct front wheel toe.

    Measured toe is: _____

    Specified toe is: _____

Problems Encountered

_____

_____

_____

Instructor's Comments

_____

_____

SUSPENSION AND STEERING JOB SHEET 20

Diagnosis of Electronically Controlled Power Steering

Name _____ Station _____ Date _____

NATEF Correlation

This Job Sheet addresses the following NATEF task:

B.19. Diagnose and adjust components of electronically controlled steering systems using a scan tool; determine necessary action.

Objective

Upon completion of this job sheet, you will be able to correctly diagnose and adjust the components of electronically controlled steering systems.

Tools and Materials

Scan tool

Service manual

Protective Clothing

Goggles or safety glasses with side shields

Describe the vehicle being worked on:

Year _____ Make _____ Model _____

VIN _____ Engine type and size _____

PROCEDURE

1. With the ignition switch on, check the electronic power steering (EPS) lamp on the instrument panel. Is the EPS lamp lit? _____ If yes, what does this indicate?

   _____

   _____

2. Start the engine; did the EPS lamp go out? _____ If yes, what does this indicate?

   _____

   _____

3. Turn the steering wheel; describe the effort it takes to turn the wheel.

   _____

   _____

4. If the EPS lamp is lit when the engine is running but it feels as though there is some power assist to the steering, turn off the ignition and remove the EPS clock fuse. Where is this fuse located?

_____

_____

5. Keep the fuse out for at least 10 seconds, and then reinstall it. Why wait 10 seconds?

_____

_____

6. Start the engine. If the EPS lamp is no longer lit, explain why not.

_____

_____

7. If the lamp is lit and the system cannot be monitored with a scan tool, locate the EPS service connector. Describe where this is.

_____

_____

8. With the ignition off, disconnect the connector and place a jumper across the designated terminals. How many terminals are at this connector?

_____

9. Turn the ignition on and count the flashes of the EPS. What did you observe?

_____

_____

10. What is indicated by the flashes?

_____

_____

11. If a scan tool can be used to retrieve diagnostic trouble codes (DTCs) from the system, connect the scan tool to the appropriate connector. Where is that connector?

_____

_____

_____

12. Retrieve any EPS trouble codes. List them.

_____

_____

13. Summarize your conclusions about this system.

_____

_____

Problems Encountered

_____

_____

_____

Instructor's Comments

_____

_____

_____

SUSPENSION AND STEERING JOB SHEET 21

Diagnosing an Electric Power Steering System

Name _____ Station _____ Date _____

NATEF Correlation

This Job Sheet addresses the following NATEF task:

B.20. Inspect and test non-hydraulic electric-power assist steering.

Objective

Upon completion of this job sheet, you will be able to inspect and test an electric power steering system and determine needed repairs.

Tools and Materials

Service manual

Scan tool

Protective Clothing

Goggles or safety glasses with side shields

Describe the vehicle being worked on:

Year _____ Make _____ Model _____

VIN _____ Engine type and size _____

Describe: the type of power steering system on the vehicle:

_____

_____

PROCEDURE

1. Obtain as much information as possible from the customer regarding when the concern is evident. What did you find out?

   _____

   _____

2. During a road test, try to duplicate the customer's concerns. Were you able to do this?

   _____

3. Watch the operation of the EPS lamp. Did it come on when the ignition switch is turned to the ON position and go off after the engine is started? What does this indicate?

   _____

   _____

4. According to the service manual, what conditions will cause the EPS lamp to light while the engine is running?

   _____

   _____

5. Inspect the system for loose connectors, broken wires and connectors, or damaged parts. What did you find?

   _____

   _____

6. Connect a scan tool to the system and retrieve all trouble codes. Did you retrieve any and what do they indicate?

   _____

   _____

7. What is indicated by the EPS lamp being lit but there are no trouble codes held in the computer's memory?

   _____

   _____

8. If the system detects a critical fault and goes into the fail-safe mode, what happens to the power steering system?

   _____

   _____

9. If the steering wheel is continuously turned back-and-forth with the vehicle stopped, the current draw of the motor will increase and cause the motor to heat up. What does the system do to prevent damage to the motor?

   _____

   _____

10. Does the system have sensor positions that must be memorized by the control unit whenever the steering gear is removed and installed, or when a sensor or EPS control unit is replaced? If yes, describe what needs to be reinitialized.

    _____

    _____

11. After diagnosing and repairing the system, clear the DTCs and road drive the vehicle. Make sure the EPS indicator does not come on. What happened?

_____

_____

Problems Encountered

_____

_____

_____

Instructor's Comments

_____

_____

_____

SUSPENSION AND STEERING JOB SHEET 22

Hybrid Vehicle Power Steering Systems

Name _____ Station _____ Date _____

NATEF Correlation

This Job Sheet addresses the following NATEF task:

   B.21. Identify hybrid vehicle power steering system electrical circuits, service and safety
   precautions.

Objective

Upon completion of this job sheet, you will be able to identify the high voltage circuits in a hybrid vehicle prior to serving the vehicle's steering system. You will also be able to identify the special procedures
required for working on this electric power steering system.

Tools and Materials

Service manual
Hybrid vehicle
Vinyl tape for insulation

Protective Clothing

Goggles or safety glasses with side shields
Insulated gloves that are dry and are not cracked, ruptured, torn, or damaged in any way.

Describe the vehicle being worked on:

Year _____ Make _____ Model _____

VIN _____ Engine type and size _____

PROCEDURE

   WARNING: *Unprotected contact with any electrically charged ("hot" or "live") high-voltage
   component could cause serious injury or death.*

1. Describe how this vehicle is identified as a hybrid.

   _____

   _____

2. After referring to the vehicle's owner's manual, briefly describe its operation.

   _____

   _____

   _____

   _____

3. How many volts are the batteries rated at?

_____

_____

4. Where are the batteries located?

_____

_____

5. How are the high-voltage cables labeled and identified?

_____

_____

6. What is used to provide short circuit protection in the high-voltage battery pack?

_____

_____

7. What isolates the high-voltage system from the rest of the vehicle when the vehicle is shut off?

_____

_____

8. Using the vehicle's service and/or owner's manual as a guide, list at least five precautions that must be adhered to when servicing this hybrid vehicle.

_____

_____

_____

_____

9. If the vehicle has been in an accident and some electrolyte from the batteries has leaked out, how should you clean up the spill?

_____

_____

_____

_____

10. Look under the hood and describe the components that are visible.

   _____

   _____

   _____

   _____

11. Without touching the high-voltage cables or components, describe the routing of the cables. Include what they appear to be connected from and to.

   _____

   _____

   _____

   _____

   WARNING: _Never assume that a hybrid vehicle is shut off simply because it is silent. Make sure the ignition key is in your pocket and not in the ignition switch._

12. Describe the procedure for totally isolating the high-voltage system from the vehicle. This often involves the removal of a service plug. Be sure to include the location of this plug in your description of the procedure.

   _____

   _____

   _____

   _____

   _____

   _____

   _____

   _____

   _____

   _____

13. Make two signs saying, _"WORKING ON HIGH VOLTAGE PARTS. DO NOT TOUCH!"_ Attach one to the steering wheel, and set the other one near the parts you are working on.

   Task Completed ☐

14. After the service plug has been removed, how long should you wait before working around or on the high-voltage system?

   _____

   _____

15. The manufacturer will also have listed certain precautions that must be followed when servicing the power steering system. Locate those in the service manual and list some of them here:

_____

_____

_____

16. Why is it important that the front wheels are perfectly straight ahead when removing and installing the steering gear assembly?

_____

_____

17. What procedures must be followed after any component related to the steering system has been removed and either reinstalled or replaced?

_____

_____

18. Does the manufacturer provide a specific procedure for centering the steering wheel? If so, briefly describe it.

_____

_____

Problems Encountered

_____

_____

_____

Instructor's Comments

_____

_____

_____

SUSPENSION AND STEERING JOB SHEET 23

Diagnosing SLA Front Suspensions

Name _____ Station _____ Date _____

NATEF Correlation

This Job Sheet addresses the following NATEF tasks:

C.1.1. Diagnose short and long-arm suspension system noises, body sway, and uneven riding height concerns; determine necessary action.

E.6. Diagnose tire pull (lead) problem; determine necessary action.

Objective

Upon completion of this job sheet, you will be able to diagnose SLA suspension system noises, body sway, and uneven riding height concerns. You will also be able to diagnose tire pull (lead) problems.

Tools and Materials

Basic hand tools   Service manual

Tape measure   Alignment machine

Protective Clothing

Goggles or safety glasses with side shields

Describe the vehicle being worked on:

Year _____ Make _____ Model _____

VIN _____ Engine type and size _____

PROCEDURE

1. Take the vehicle for a road test and note any abnormal or questionable noises and handling characteristics. Record the results of the check.

   _____

   _____

   _____

Noises

2. Determine the cause of the abnormal noises using these guidelines.

   a. Tire noise varies with road surface conditions, whereas differential noise is not affected when various road surfaces are encountered.

b. Uneven tread surfaces may cause tire noises that seem to originate elsewhere in the vehicle. These noises may be confused with differential noise. Differential noise usually varies with acceleration and deceleration, whereas tire noise remains more constant in relation to these forces. Tire noise is most pronounced on smooth asphalt road surfaces at speeds of 15 to 45 mph.

c. Rattling on road irregularities can be caused by worn shock absorber bushings or grommets, worn spring insulators, a broken coil spring or spring insulators, worn control arm bushings, worn stabilizer bar bushings, worn leaf spring shackles and bushings, or worn torsion bars, anchors, and bushings.

d. Dry or worn control arm bushings may cause a squeaking noise on irregular road surfaces.

3. What did you determine to be the cause of the noises observed during the road test?

_____

_____

_____

Curb Riding Height

4. a. The curb riding height is determined mainly by spring condition. Other suspension components, such as control arm bushings, affect curb riding height if they are worn. Since incorrect curb riding height affects most of the other suspension angles, this measurement is critical. Reduced curb riding height on the front suspension may cause decreased directional stability. If the curb riding height is reduced on one side of the front suspension, the steering may pull to one side. Reduced rear suspension height increases steering effort and causes rapid steering wheel return after turning a corner. Harsh riding occurs when the curb riding height is less than specified. The curb riding height must be measured at the vehicle manufacturer's specified location, which varies depending on the type of suspension system. The following is a typical procedure.

   (1) Park the vehicle on a level floor or on an alignment rack and measure the curb riding height from the floor to the center of the lower control arm mounting bolt on both sides of the front suspension. What was your measurement?

   _____

   (2) Measure the rear suspension from the floor to the center of the strut rod mounting bolt. What was your measurement?

   _____

   (3) Compare your measurements to specifications and summarize your conclusions.

   _____

   _____

b. If the curb riding height is less than specified, the control arms and bushings should be inspected and replaced. When the control arms and bushings are in normal condition, the reduced curb riding height may be caused by sagged springs or worn-out air shocks.

c. If the curb ride height is not within specifications, what is the likely cause?

_____

_____

Component Inspection

5. a. Based on the road test, list all abnormal or questionable handling behaviors of the vehicle. Describe them here.

_____

_____

_____

b. Use the following to diagnose the cause of the handling concerns.

Steering Pull

(1) A vehicle should maintain straight-ahead forward direction on smooth, straight road surfaces without excessive steering wheel correction by the driver. If the steering gradually pulls to one side on a smooth, straight road surface, a tire, steering, or suspension defect is present. Check for tires of different types, sizes, designs, or inflation pressures on opposite sides of the vehicle.

(2) Check for incorrect front suspension alignment angles. Check the condition of the upper and lower control arm bushings; if worn, these can affect the caster and camber angles on the front suspension.

(3) Check for low ride height on one side of the front suspension.

(4) Check for a broken center bolt in leaf springs. This problem allows one side of the front suspension to move rearward in relation to the other side.

Excessive Chassis Oscillations

Check for worn or leaking shock absorbers.

Excessive Sway on Road Irregularities

(1) Check for a weakened or disconnected stabilizer bar.

(2) Check for worn stabilizer bar bushings.

Reduced Directional Stability

(1) Check for weak or broken coil springs that affect front suspension alignment angles, which may cause reduced directional stability.

(2) Check for weak or broken leaf springs.

(3) Check for weak torsion bars or worn bushings or anchors.

Erratic Steering on Irregular Road Surfaces

Inspect the upper and lower control arms for cracks, distortion, and worn bushings.

Increased Steering Effort

Check for low ride height at the rear suspension.

Rapid Steering Wheel Return

Check for low ride height at the rear suspension.

Harsh Riding

(1) Check for low ride height at the front and rear suspension.

(2) Check for weak or broken coil springs.

(3) Check for weak or broken leaf springs.

(4) Check for worn torsion bar components such as pivot cushion bushings.

(5) Check the control arm bushings for wear or damage.

c. Based on the guidelines above and the road test, summarize the needed repairs.

_____

_____

_____

Problems Encountered

_____

_____

_____

Instructor's Comments

_____

_____

_____

SUSPENSION AND STEERING JOB SHEET 24

Diagnosing Strut-Type Front Suspensions

Name _____ Station _____ Date _____

NATEF Correlation

This Job Sheet addresses the following NATEF task:

C.1.2. Diagnose strut suspension system noises, body sway, and uneven riding height concerns; determine necessary action.

Objective

Upon completion of this job sheet, you will be able to diagnose strut suspension system noises, body sway, and uneven riding height concerns.

Tools and Materials
Basic hand tools
Tape measure
Service manual
Alignment machine

Protective Clothing
Goggles or safety glasses with side shields

Describe the vehicle being worked on:
Year _____ Make _____ Model _____

VIN _____ Engine type and size _____

PROCEDURE

1. Take the vehicle for a road test and note any abnormal or questionable noises and handling characteristics. Record the results of the check.

   _____

   _____

Noises
a. Determine the cause of the abnormal noises using these guidelines.
   (1) Tire noise varies with road surface conditions, whereas differential noise is not affected when various road surfaces are encountered.
   (2) Uneven tread surfaces may cause tire noises that seem to originate elsewhere in the vehicle. These noises may be confused with differential noise. Differential noise usually varies with acceleration and deceleration, whereas tire noise remains more constant in relation to

these forces. Tire noise is most pronounced on smooth asphalt road surfaces at speeds of 15 to 45 mph.

(3) Rattling on road irregularities can be caused by worn strut bushings or grommets, worn spring insulators, a broken coil spring or spring insulators, worn control arm bushings, worn stabilizer bar bushings, and worn strut rod grommets.

(4) Chatter during cornering can be caused by worn upper strut mounts.

(5) Dry or worn control arm bushings may cause a squeaking noise on irregular road surfaces.

b. What did you determine to be the cause of the noises observed during the road test?

_____

_____

_____

Curb Riding Height

a. The curb riding height is determined mainly by spring condition. Other suspension components, such as control arm bushings, affect curb riding height if they are worn. Since incorrect curb riding height affects most of the other suspension angles, this measurement is critical. Reduced curb riding height on the front suspension may cause decreased directional stability. If the curb riding height is reduced on one side of the front suspension, the steering may pull to one side. Reduced rear suspension height increases steering effort and causes rapid steering wheel return after turning a corner. Harsh riding occurs when the curb riding height is less than specified. The curb riding height must be measured at the vehicle manufacturer's specified location, which varies depending on the type of suspension system. The following is a typical procedure.

(1) Park the vehicle on a level floor or on an alignment rack and measure the curb riding height from the floor to the center of the lower control arm mounting bolt on both sides of the front suspension. What was your measurement?

_____

(2) Measure the rear suspension from the floor to the center of the strut rod mounting bolt. What was your measurement?

_____

(3) Compare your measurements to specifications and summarize your conclusions.

_____

_____

b. If the curb riding height is less than specified, the control arms and bushings should be inspected and replaced. When the control arms and bushings are in normal condition, the reduced curb riding height may be caused by worn-out or damaged struts.

c. If the curb riding height is not within specifications, what is the likely cause?

_____

_____

Component Inspection

a. Based on the road test, list all abnormal or questionable handling behaviors of the vehicle. Describe them here.

_____

_____

_____

b. Use the following guidelines to diagnose the cause of the handling concerns.

Steering Pull

(1) A vehicle should maintain straight-ahead forward direction on smooth, straight road surfaces without excessive steering wheel correction by the driver. If the steering gradually pulls to one side on a smooth, straight road surface, a tire, steering, or suspension defect is present. Check for tires of different types, sizes, designs, or inflation pressures on opposite sides of the vehicle.

(2) Check for incorrect front suspension alignment angles. Check the condition of the lower control arm bushings and strut bearings; if worn, these can affect the caster and camber angles on the front suspension.

(3) Check for low riding height on one side of the front suspension.

Excessive Chassis Oscillations

Check for worn or leaking struts.

Excessive Sway on Road Irregularities

(1) Check for a weakened or disconnected stabilizer bar.

(2) Check for worn stabilizer bar bushings.

Reduced Directional Stability

Check for weak or broken coil springs that affect front suspension alignment angles; this problem may cause reduced directional stability.

Erratic Steering on Irregular Road Surfaces

Inspect the lower control arms for cracks, distortion, and worn bushings.

Increased Steering Effort

Check for low riding height at the rear suspension.

Rapid Steering Wheel Return

Check for low riding height at the rear suspension.

Harsh Riding

(1) Check for low riding height at the front and rear suspension.

(2) Check for weak or broken coil springs.

(3) Check the control arm bushings for wear or damage.

2. Based on the guidelines above and the road test, summarize the needed repairs.

_____

_____

_____

Problems Encountered

_____

_____

_____

Instructor's Comments

_____

_____

_____

SUSPENSION AND STEERING JOB SHEET 25

Servicing Coil Springs and Control Arm Bushings

Name _____ Station _____ Date _____

NATEF Correlation

This Job Sheet addresses the following NATEF task:

   C.1.3. Remove, inspect, and install upper and lower control arms, bushings, shafts, and rebound bumpers.

Objective

Upon completion of this job sheet, you will be able to remove, inspect, and install upper and lower arms, bushings, shafts, and rebound bumpers.

Tools and Materials

Floor jack	Torque wrench
Lift	Spring compressing tool
Hand tools	Safety stands
Bushing removal and installation tools	Ball joint spreader tool

Protective Clothing

Goggles or safety glasses with side shields

Describe the vehicle being worked on:

Year _____ Make _____ Model _____

VIN _____ Engine type and size _____

PROCEDURE

   WARNING: *During control arm and spring replacement, the coil spring tension is supported by a compressing tool or floor jack. Always follow the vehicle manufacturer's recommended control arm and spring replacement procedures very carefully. Serious personal injury or property damage may occur if the spring tension is released suddenly.*

   NOTE: *Follow the procedure below when a lower control arm and/or spring is replaced on a vehicle with an SLA suspension system in which the coil springs are positioned between the lower control arm and the frame.*

1. Lift the vehicle on a hoist until the tires are a short distance off the floor, and allow the front suspension to drop downward. An alternate method is to lift the vehicle with a floor jack, and then support the chassis securely on safety stands so the front suspension drops downward.

   Task Completed ☐

2. Disconnect the lower end of the shock absorber.

   Task Completed ☐

3. Disconnect the stabilizer bar from the lower control arm.   Task Completed ☐

4. Install a spring compressor and turn the spring compressor bolt until the   Task Completed ☐
   spring is compressed. Make sure all the spring tension is supported by the
   compressing tool.

5. Place a floor jack under the lower control arm, and raise the jack until the   Task Completed ☐
   control arm is raised and the rebound bumper is not making contact.

6. Remove the lower ball joint cotter pin and nut, and use a threaded expan-   Task Completed ☐
   sion tool to loosen the lower ball joint stud.

7. Lower the floor jack very slowly to lower the control arm and coil spring.   Task Completed ☐

8. Disconnect the lower control arm inner mounting bolts, and remove the   Task Completed ☐
   lower control arm.

9. Rotate the compressing tool bolt to release the spring tension, and remove   Task Completed ☐
   the spring from the control arm.

10. Inspect the lower control arm for a bent condition or cracks. If either of   Task Completed ☐
    these conditions is present, replace the control arm.

11. Visually inspect the upper and lower spring insulators for cracks and wear,
    and inspect the spring seat areas in the chassis and lower control arm.
    Worn or cracked spring insulators must be replaced. Describe your
    findings.

    _____

    _____

12. Visually inspect all control arm bushings. Loose or worn bushings must be
    replaced. Describe your findings.

    _____

    _____

13. To remove an old bushing, install the bushing removal or receiver tool on   Task Completed ☐
    the control arm over the bushing. Tighten the attachment nuts until the
    tool is securely in place. Then install the plate bolt through the plate and
    into the receiver tool.

14. Place the remover tool into position on the bushing. Install the nut onto   Task Completed ☐
    the plate bolt. Remove the bushing from the control arm by turning the
    bolt.

PROCEDURE (INSTALLATION)

1. With the vehicle up on the hoist, install the bushing on the bolt. Position   Task Completed ☐
   it onto the housing. Align the bushing installer arrow with the arrow on
   the receiver for proper indexing of the bushing. Install the nut onto the
   bolt.

2. Press the bushing into the control arm by turning the bolt.   Task Completed ☐

     a. Is the end flange flush against the face of the control arm?   Yes ☐   No ☐

     b. If no, correct.   Task Completed ☐

3. Use a screw-type jack stand to position the control arm into the bracket. Install the bolt and nut. (Do not torque.)   Task Completed ☐

4. Install the brake-line bracket to the frame. Torque the screw to specifications.   Task Completed ☐

     Torque specification _____

5. Reconnect the brake cables to the bracket and reinstall the brake cable to the hook. Adjust the cable as necessary.   Task Completed ☐

6. Support the vehicle at curb height. Tighten the control arm bolt to specifications.   Task Completed ☐

     CAUTION: *Do not torque the control arm bolt without having the suspension at curb height. The torsional bushings will wear prematurely if torqued at improper height.*

     Torque specification _____

7. Remove the jack stands. Install the wheel assembly.   Task Completed ☐

     Lug nut torque _____

8. Lower the vehicle from the hoist.   Task Completed ☐

Problems Encountered

_____

_____

_____

Instructor's Comments

_____

_____

_____

SUSPENSION AND STEERING JOB SHEET 26

Remove and Replace a Strut Rod Assembly

Name _____ Station _____ Date _____

NATEF Correlation

This Job Sheet addresses the following NATEF task:

C.1.4. Remove, inspect, install, and adjust strut rods (compression/tension) and bushings.

Objective

Upon completion of this job sheet, you will be able to remove, inspect, install, and adjust strut rods and bushings properly.

Tools and Materials

Frame hoist	Fender covers
Hand tools	Service manual
Torque wrench	

Protective Clothing

Goggles or safety glasses with side shields

Describe the vehicle being worked on:

Year _____ Make _____ Model _____

VIN _____ Engine type and size _____

Describe the type of front suspension:

_____

Describe the type of rear suspension:

_____

What strut are you going to remove and replace?

_____

PROCEDURE

1. Raise the hood and place fender covers on the vehicle.   Task Completed ☐

2. Raise the vehicle on the frame hoist.   Task Completed ☐

3. Remove the wheel and tire from the vehicle.   Task Completed ☐

4. Disconnect any brake lines or wires attached to the strut assembly.   Task Completed ☐

5. Disconnect steering linkage from the strut assembly, if so designed.   Task Completed ☐

6. Remove the bolts that connect the strut assembly to the spindle.

Task Completed ☐

7. Lower the vehicle for access to the upper bolts. Do not lower it all the way to the ground.

Task Completed ☐

8. Remove the nuts mounting the strut assembly to the strut tower.

Task Completed ☐

NOTE: *Do not remove the large center nut holding the strut and the mounting bracket together. When you remove the mounting nuts, do not allow the strut to fall out of the car.*

9. Carefully remove the strut assembly from the vehicle. Do not allow it to hang on any other component of the vehicle.

Task Completed ☐

10. To reinstall the strut assembly, reverse the preceding procedure using the service manual for the proper torque specifications of all mounting nuts and bolts.

Task Completed ☐

Problems Encountered

_____

_____

_____

Instructor's Comments

_____

_____

_____

SUSPENSION AND STEERING JOB SHEET 27

Servicing Upper and Lower Ball Joints

Name _____ Station _____ Date _____

NATEF Correlation

This Job Sheet addresses the following NATEF task:

C.1.5. Remove, inspect, and install upper and lower ball joints.

Objective

Upon completion of this job sheet, you will be able to remove, inspect and install the upper and lower ball joints on SLA suspension systems.

Tools and Materials

Hydraulic floor jack   Ball joint loosening tool

Safety stands   Torque wrench

Dial indicator

Protective Clothing

Goggles or safety glasses with side shields

Describe the vehicle being worked on:

Year _____ Make _____ Model _____

VIN _____ Engine type and size _____

PROCEDURE

1. Raise the front of the vehicle with a floor jack lift pad positioned on the specified lifting point.

   Task Completed ☐

2. Install safety stands near the outer ends of the lower control arms. Lower the floor jack so the control arms are supported on the safety stands. Remove the floor jack.

   Task Completed ☐

3. Attach a dial indicator for ball joint measurement to the lower control arm, and position the dial indicator stem against the lower end of the steering knuckle next to the ball joint retaining nut if the suspension has a lower compression-loaded ball joint. If the suspension has a tension-loaded ball joint, place the indicator stem against the top of the ball joint stud. What type of ball joint is it and where should you place the dial indicator's stem?

   _____

   _____

4. Preload the dial indicator stem 0.250 in (6.35 mm), and zero the dial indicator.

   Task Completed ☐

5. Place a pry bar under the front tire and lift straight upward on the pry bar while a co-worker observes the dial indicator. Task Completed ☐

6. Compare the reading on the dial indicator to the vehicle manufacturer's specifications. What are the specifications?

   _____

   _____

7. Summarize your service recommendations.

   _____

   _____

   _____

   _____

8. Make sure the front wheel bearings are properly adjusted. Then attach the dial indicator to the lower control arm, and position the dial indicator against the inner edge of the wheel rim. Preload the dial indicator 0.250 in (6.35 mm) and place the dial in the zero position. Task Completed ☐

9. Grasp the tire at the top and bottom, and try to rock the tire inward and outward while a co-worker observes the dial indicator. Task Completed ☐

10. Compare the reading on the dial indicator to the vehicle manufacturer's specifications. What are the specifications?

    _____

    _____

11. Summarize your service recommendations.

    _____

    _____

    _____

    _____

12. Repeat the measurements in steps 3 through 11 on the opposite side of the front suspension. Task Completed ☐

13. Based on your ball joint measurements, state all the necessary ball joint service and explain the reasons for your diagnosis.

    _____

    _____

    _____

    _____

14. To replace a ball joint, remove the wheel cover and loosen the wheel nuts. Task Completed ☐

15. Lift the vehicle with a floor jack and place safety stands under the chassis so the front suspension is allowed to drop downward. Lower the vehicle onto the safety stands and remove the floor jack.

Task Completed ☐

16. Remove the wheel and place a floor jack under the outer end of the lower control arm. Operate the floor jack and raise the lower control arm until the ball joints are unloaded.

Task Completed ☐

17. Remove other components, such as the brake caliper, rotor, and drive axle, as required to gain access to the ball joints. What did you need to remove?

_____

_____

18. Remove the cotter pin in the ball joint or joints that require replacement, and loosen, but do not remove, the ball joint stud nuts.

Task Completed ☐

19. Loosen the ball joint stud tapers in the steering knuckle. A threaded expansion tool is available for this purpose.

Task Completed ☐

20. Remove the ball joint nut and lift the knuckle off the ball joint stud. Block or tie up the knuckle and hub assembly to gain access to the ball joint.

Task Completed ☐

21. If the ball joint is riveted to the control arm, drill and punch out the rivets and bolt the new ball joint to the control arm.

Task Completed ☐

22. If the ball joint is pressed into the lower control arm, remove the ball joint dust boot and use a pressing tool to remove and replace the ball joint.

Task Completed ☐

23. If the ball joint housing is threaded into the control arm, use the proper size socket to remove and install the ball joint. The replacement ball joint must be torqued to the manufacturer's specifications. If a minimum of 125 foot-pounds of torque cannot be obtained, the control arm threads are damaged and control arm replacement is necessary.

Task Completed ☐

24. If the ball joint is bolted to the lower control arm, install the new ball joint and tighten the bolt and nuts to the specified torque. What is the specified torque? _____

25. Clean and inspect the ball joint stud tapered opening in the steering knuckle. If this opening is out-of-round or damaged, the knuckle must be replaced. What did you find?

_____

_____

26. Check the fit of the ball joint stud in the steering knuckle opening. This stud should fit snugly in the opening and only the threads on the stud should extend through the knuckle. If the ball joint stud fits loosely in the knuckle tapered opening, either this opening is worn or the wrong ball joint has been supplied. What did you find?

_____

_____

27. Install the ball joint stud in the steering knuckle opening, making sure the stud is straight and centered. Install the stud nut and tighten this nut to the specified torque. Install a new cotter pin through the stud and nut. Do not loosen the nut to align the nut and stud openings. What is the specified torque? _____

28. Reassemble the components that were removed in steps 16, 17, and 18. Make sure the wheel nuts are tightened to the specified torque. What is the specified torque? _____

Problems Encountered

_____

_____

_____

Instructor's Comments

_____

_____

_____

SUSPENSION AND STEERING JOB SHEET 28

Removing a Steering Knuckle

Name _____ Station _____ Date _____

NATEF Correlation

This Job Sheet addresses the following NATEF task:

C.1.6. Remove, inspect, and install steering knuckle assemblies.

Objective

Upon completion of this job sheet, you will be able to remove, inspect, and install a steering knuckle assembly on a vehicle.

Tools and Materials

Floor jack   Torque wrench

Safety stands   Driving tools

Tire-rod end puller

Protective Clothing

Goggles or safety glasses with side shields

Describe the vehicle being worked on:

Year _____ Make _____ Model _____

VIN _____ Engine type and size _____

PROCEDURE

1. Remove the wheel cover and loosen the front wheel nuts and the drive axle nut.   Task Completed ☐

2. Lift the vehicle chassis on a hoist and allow the front suspension to drop downward. Remove the front wheel, brake caliper, brake rotor, and drive axle nut. Tie the brake caliper to a suspension component; do not allow the caliper to hang on the end of the brake hose.   Task Completed ☐

3. Remove the inner end of the drive axle from the transaxle with a pulling or prying action.   Task Completed ☐

4. Remove the outer end of the drive axle from the steering knuckle and hub. On some models, a puller is required for this operation.   Task Completed ☐

5. Make sure the vehicle's weight is supported on the hoist with the front suspension dropped downward. Remove the mounting bolts in the lower control arm.   Task Completed ☐

6. Remove the cotter pin from the tie-rod nut.   Task Completed ☐

7. Remove the outer tie-rod nut and use a puller to disconnect the tie-rod end from the steering knuckle.   Task Completed ☐

8. If an eccentric cam is used on one of the strut-to-knuckle bolts, mark the cam and bolt position in relation to the strut and remove the strut-to-knuckle bolts.   Task Completed ☐

9. Remove the knuckle from the strut and lift the knuckle out of the chassis.   Task Completed ☐

10. Pry the dust deflector from the steering knuckle with a large flat-blade screwdriver.   Task Completed ☐

11. Use a puller to remove the ball joint from the steering knuckle.   Task Completed ☐

12. Check the ball joint and tie-rod end openings in the knuckle for wear and out-of-round. Replace the knuckle if these openings are worn or out-of-round. Describe the condition of ball joint opening and the tie-rod end opening in the knuckle.

_____

_____

13. State the necessary steering knuckle and related repairs, and explain the reason for your diagnosis.

_____

_____

_____

_____

14. Install the ball joint into the steering knuckle.   Task Completed ☐

15. Use the proper driving tool to reinstall the dust deflector in the steering knuckle.   Task Completed ☐

16. Fit the knuckle onto the strut.   Task Completed ☐

17. If an eccentric cam is used on one of the strut-to-knuckle bolts, align the eccentric and bolt to the marks made previously, then install and tighten the bolts.   Task Completed ☐

18. Install the outer tie-rod end into the steering knuckle and tighten the nut. Task Completed ☐

19. Install a new cotter pin in the tie-rod nut.   Task Completed ☐

20. Install the mounting bolts in the lower control arm.   Task Completed ☐

21. Install the outer end of the drive axle into the steering knuckle and hub.   Task Completed ☐

22. Install the inner end of the drive axle into the transaxle.   Task Completed ☐

23. Attach the brake rotor and then the caliper assembly.   Task Completed ☐

24. Install the front wheel and drive axle nut.   Task Completed ☐

25. Lower the vehicle.   Task Completed ☐

26. Tighten the lug nuts and drive axle nut to specifications. Then install the wheel cover. What are the specifications?

_____

_____

Problems Encountered

_____

_____

_____

Instructor's Comments

_____

_____

_____

SUSPENSION AND STEERING JOB SHEET 29

Servicing Coil Springs and Lower Control Arm

Name _____ Station _____ Date _____

NATEF Correlation

This Job Sheet addresses the following NATEF task:

C.1.7. Remove, inspect, and install short and long-arm suspension system coil springs and spring insulators.

Objective

Upon completion of this job sheet, you will be able to remove, inspect, and install SLA suspension system coil springs and spring insulators.

Tools and Materials

Floor jack   Safety stands

Spring compressing tool   Ball joint spreader tool

Protective Clothing

Goggles or safety glasses with side shields

Describe the vehicle being worked on:

Year _____ Make _____ Model _____

VIN _____ Engine type and size _____

PROCEDURE

WARNING: *During control arm and spring replacement, the coil spring tension is supported by a compressing tool or floor jack. Always follow the vehicle manufacturer's recommended control arm and spring replacement procedures very carefully. Serious personal injury or property damage may occur if the spring tension is released suddenly.*

NOTE: *Follow this procedure when a lower control arm and/or spring is replaced on a vehicle with a short-and-long arm suspension system with the coil springs positioned between the lower control arm and the frame.*

1. Lift the vehicle on a hoist until the tires are a short distance off the floor, and allow the front suspension to drop downward. An alternate method is to lift the vehicle with a floor jack, and then support the chassis securely on safety stands so the front suspension drops downward.   Task Completed ☐

2. Disconnect the lower end of the shock absorber.   Task Completed ☐

3. Disconnect the stabilizer bar from the lower control arm.   Task Completed ☐

4. Install a spring compressor and turn the spring compressor bolt until the spring is compressed. Make sure all the spring tension is supported by the compressing tool.

Task Completed ☐

5. Place a floor jack under the lower control arm, and raise the jack until the control arm is raised and the rebound bumper is not making contact.

Task Completed ☐

6. Remove the lower ball joint cotter pin and nut, and use a threaded expansion tool to loosen the lower ball joint stud.

Task Completed ☐

7. Lower the floor jack very slowly to lower the control arm and coil spring.

Task Completed ☐

8. Disconnect the lower control arm inner mounting bolts, and remove the lower control arm.

Task Completed ☐

9. Rotate the compressing tool bolt to release the spring tension, and remove the spring from the control arm.

Task Completed ☐

10. Inspect the lower control arm for a bent condition or cracks. If either of these conditions is present, replace the control arm.

Task Completed ☐

11. Visually inspect all control arm bushings. Loose or worn bushings must be replaced. Describe your findings.

_____

_____

12. Visually inspect the upper and lower spring insulators for cracks and wear, and inspect the spring seat areas in the chassis and lower control arm. Worn or cracked spring insulators must be replaced. Describe your findings.

_____

_____

13. Reverse steps 1 through 9 to install the lower control arm. Be sure the coil spring and insulators are properly seated in the lower control arm and in the upper spring seat.

Task Completed ☐

Problems Encountered

_____

_____

_____

Instructor's Comments

_____

_____

_____

SUSPENSION AND STEERING JOB SHEET 30

Torsion Bar Adjustment

Name _____ Station _____ Date _____

NATEF Correlation

This Job Sheet addresses the following NATEF task:

C.1.8. Remove, inspect, install, and adjust suspension system torsion bars; inspect mounts.

Objective

Upon completion of this job sheet, you will be able to inspect, install, and adjust suspension system torsion bars and to inspect the mounts.

Tools and Materials

Measuring tape

Lift

Protective Clothing

Goggles or safety glasses with side shields

Describe the vehicle being worked on:

Year _____ Make _____ Model _____

VIN _____ Engine type and size _____

PROCEDURE

1. Check all of the components of the suspension and steering system for things that may affect curb riding height. List all problems found and suggest what must be done to correct them.

   _____

   _____

2. On torsion bar front suspension systems, the torsion bars may be adjusted to correct the curb riding height. The curb riding height is measured with the vehicle on a lift and the tires supported on the lift.

   Task Completed ☐

3. Measure the distance from the center of the front lower control arm bushing to the lift. Your measurement was:

   _____

   _____

4. Then measure the distance from the lower end of the front spindle to the lift. Your measurement was:

_____

_____

5. The difference between these two readings is the curb riding height. What is the curb riding height?

_____

_____

6. If the curb riding height is not correct on a torsion bar front suspension, the torsion bar anchor adjusting bolts must be rotated until the curb riding height equals the vehicle manufacturer's specifications. What are the specifications?

_____

_____

Problems Encountered

_____

_____

_____

Instructor's Comments

_____

_____

_____

SUSPENSION AND STEERING JOB SHEET 31

Servicing Stabilizer Bars

Name _____ Station _____ Date _____

NATEF Correlation

This Job Sheet addresses the following NATEF task:

C.1.9. Remove, inspect, and install stabilizer bar bushings, brackets, and links.

Objective

Upon completion of this job sheet, you will be able to correctly remove, inspect, and install stabilizer bar bushings, brackets, and links.

Tools and Materials

Lift

Hand tools

Torque wrench

Bushing tools

Protective Clothing

Goggles or safety glasses with side shields

Describe the vehicle being worked on:

Year _____ Make _____ Model _____

VIN _____ Engine type and size _____

PROCEDURE

1. Lift the vehicle on a hoist and allow both sides of the front suspension to drop downward as the vehicle chassis is supported on the hoist.

   Task Completed ☐

2. Remove the mounting bolts at the outer ends of the stabilizer bar and remove the bushings, grommets, brackets, or spacers.

   Task Completed ☐

3. Remove the mounting bolts in the center area of the stabilizer bar.

   Task Completed ☐

4. Remove the stabilizer bar from the chassis.

   Task Completed ☐

5. Visually inspect all stabilizer bar components, such as bushings, bolts, and spacer sleeves. Replace the stabilizer bar, grommets, bushings, brackets, or spacers as required. Split bushings may be removed over the stabilizer bar. Bushings that are not split must be pulled from the bar. Summarize your findings.

   _____

   _____

6. Reverse steps 2 through 4 to install the stabilizer bar. Make sure all stabilizer bar components are installed in the original position, and tighten all fasteners to the specified torque. What is the specified torque?

_____

7. Some vehicle manufacturers specify that stabilizer bars must have equal distances between the outer bar ends and the lower control arms. Always refer to the manufacturer's recommended measurement procedure. If this measurement is required, adjust the nut on the outer stabilizer bar mounting bolt until equal distances are obtained between the outer bar ends and the lower control arms. Worn grommets can cause these distances to be unequal. Did you need to do this? If so, what did you do?

_____

_____

_____

Problems Encountered

_____

_____

_____

Instructor's Comments

_____

_____

_____

SUSPENSION AND STEERING JOB SHEET 32

Replace the Cartridge in a Strut Suspension

Name _____ Station _____ Date _____

NATEF Correlation

This Job Sheet addresses the following NATEF task:

C.1.10. Remove, inspect, and install strut cartridge or assembly, strut coil spring, insulators (silencers), and upper strut bearing mount.

Objective

Upon completion of this job sheet, you will be able to remove, inspect, and install a strut cartridge or assembly, as well as strut coil springs, insulators (silencers), and upper strut bearing mounts.

Tools and Materials

Floor jack   Pipe cutter

Safety stands   Torque wrench

Coil spring compressor

Protective Clothing

Goggles or safety glasses with side shields

Describe the vehicle being worked on:

Year _____ Make _____ Model _____

VIN _____ Engine type and size _____

PROCEDURE

1. With the vehicle parked on the shop floor, perform a strut bounce test. Based on the bounce test results, state the strut condition and give the reason for your diagnosis.

   _____

   _____

   _____

2. Raise the vehicle on a hoist or with a floor jack. If a floor jack is used to raise the vehicle, lower the vehicle onto safety stands placed under the chassis so that the lower control arms and front wheels drop downward. Remove the floor jack from under the vehicle.   Task Completed ☐

3. Remove the brake line and antilock brake system (ABS) wheel-speed sensor wire from clamps on the strut. In some cases, these clamps may have to be removed from the strut.   Task Completed ☐

4. Punch mark the cam bolt in relation to the strut, remove the strut to steering knuckle retaining bolts, and remove the strut from the knuckle.

Task Completed ☐

5. Remove the upper strut mounting bolts on top of the strut tower; remove the strut and spring assembly.

Task Completed ☐

   CAUTION: *Always use a coil spring compressing tool according to the tool or vehicle manufacturer's recommended service procedure. Be sure the tool is properly installed on the spring. If a coil spring slips off the tool when the spring is compressed, severe personal injury or property damage may occur.*

   CAUTION: *Never loosen the upper strut mount retaining nut on the end of the strut rod unless the spring is compressed enough to remove all spring tension from the upper strut mount. If this nut is loosened with spring tension on the upper mount, this mount becomes a very dangerous projectile that may cause serious personal injury or property damage.*

   WARNING: *Never clamp the lower strut or shock absorber chamber in a vise with excessive force. This action may distort the lower chamber and affect piston movement in the strut or shock absorber.*

6. Install the spring compressing tool on the coil spring according to the tool or vehicle manufacturer's recommended procedure. If the coil spring has an enamel-type coating and the compressing tool contacts the coil spring, tape the spring where the compressing tool contacts the spring.

Task Completed ☐

7. Turn the nut on top of the compressing tool until all the spring tension is removed from the upper strut mount.

Task Completed ☐

8. Install a bolt and two nuts in the upper strut to knuckle mounting bolt holes. Install a nut on each side of the strut flange. Clamp this bolt securely in a vise to hold the strut and coil assembly and the compressing tool.

Task Completed ☐

9. Use the bar on the spring compressing tool to keep the strut and spring assembly from turning, and loosen the nut on the upper strut mount. Be sure all the spring tension is removed from the upper strut mount before loosening this nut.

Task Completed ☐

10. Remove the nut, upper strut mount, upper insulator, coil spring, spring bumper, and lower insulator.

Task Completed ☐

11. Inspect the strut, upper strut mount, coil spring, spring insulators, and spring bumper. Based on this inspection, list the necessary strut and spring service, and give the reasons for your diagnosis.

_____

_____

_____

12. Install a bolt and two nuts into the *upper strut to knuckle mounting bolt hole*. Place a nut on the inside and outside of the strut flange.

Task Completed ☐

13. Clamp this bolt in a vise to hold the strut.

Task Completed ☐

14. Locate the line groove near the top of the strut body, and use a pipe cutter installed in this groove to cut the top of the strut body.

15. After the cutting procedure, remove the strut piston assembly from the strut.

16. Remove the strut from the vise and dump the oil from the strut.

17. Place the special tool supplied by the vehicle manufacturer or cartridge manufacturer on top of the strut body and strike the tool with a plastic hammer until the tool shoulder contacts the top of the strut body. This action removes burrs from the strut body and places a slight flare on the body.

18. Remove the tool from the strut body.

19. Place the new cartridge in the strut body and turn the cartridge until it settles into indentations in the bottom of the strut body.

20. Place the new nut over the cartridge.

21. Using a special tool supplied by the vehicle or cartridge manufacturer, tighten the nut to the specified torque. What is the specified strut nut torque?

_____

22. Move the strut piston rod in and out several times to check for proper strut operation. What did you find?

_____

_____

23. Install a bolt in the upper strut to knuckle retaining bolt, and clamp this bolt in a vise to hold the strut, spring, and compressing tool as in the disassembly procedure.

24. Install the lower insulator on the lower strut spring seat and be sure the insulator is properly seated.

25. Install the spring bumper on the strut rod.

26. With the coil spring compressed in the spring compressing tool, install the spring on the strut. Be sure the spring is properly seated on the lower insulator spring seat.

27. Be sure the strut piston rod is fully extended and install the upper insulator on top of the coil spring.

28. Install the upper strut mount on the upper insulator.

29. Be sure the spring, upper insulator, and upper strut mount are properly positioned and seated on the coil spring and strut piston rod.

30. Use the compressing tool bar to keep the strut and spring from turning, then tighten the strut piston rod nut to the specified torque. What is the specified strut piston rod nut torque? _____

31. Rotate the upper strut mount until the lowest bolt in this mount is aligned with the tab on the lower spring seat.

32. Gradually loosen the nut on the compressing tool until all the spring tension is released from the tool, and remove the tool from the spring.

Task Completed ☐

33. Install the strut and spring assembly with the upper strut mounting bolts extending through the bolt holes in the strut tower. Tighten the nuts on the upper strut mounting bolts to the specified torque. What is the specified upper strut mount nut torque? _____

34. Install the lower end of the strut in the steering knuckle to the proper depth. Align the punch marks on the cam bolt and strut that were placed on these components during disassembly, and tighten the strut-to-knuckle retaining bolts to the specified torque. What is the specified strut-to-knuckle bolt torque? _____

35. Install the brake hose in the clamp on the strut. Place the ABS wheel-speed sensor wire in the strut clamp if the vehicle is equipped with ABS.

Task Completed ☐

36. Raise the vehicle with a floor jack, remove the safety stands, and lower the vehicle onto the shop floor.

Task Completed ☐

Problems Encountered

_____

_____

_____

Instructor's Comments

_____

_____

_____

SUSPENSION AND STEERING JOB SHEET 33

Lubricate the Chassis

Name _____ Station _____ Date _____

NATEF Correlation

This Job Sheet addresses the following NATEF task:

C.1.11. Lubricate suspension and steering systems.

Objective

Upon completion of this job sheet, you will be able to inspect and lubricate the suspension and steering systems on a vehicle.

Tools and Materials

Lift

Shop towels

Manual or pneumatic grease gun

Service manual

Zerk fittings

Protective Clothing

Goggles or safety glasses with side shields

Describe the vehicle being worked on:

Year _____ Make _____ Model _____

VIN _____ Engine type and size _____

PROCEDURE

1. Refer to the service manual for the lubrication points on the steering and suspension system of the vehicle. How many lubrication points are identified and where are they?

   _____

   _____

   _____

2. Safely raise the vehicle and lock the lift or set the vehicle on safety stands.   Task Completed ☐

3. Locate all of the lubrication points and wipe the fittings clean with a shop towel.   Task Completed ☐

4. How many of the lubrication points do not have a Zerk fitting but have a plug?

   _____

5. Remove the plugs and install new Zerk fittings.   Task Completed ☐

6. Carefully look at the joints and determine if the joint boots are sealed or not. Record your findings here.

_____

7. Push the grease gun nozzle straight onto a Zerk fitting and pump grease slowly into the joint. If the joint has a sealed boot, put just enough grease into the joint to cause the boot to expand slightly. If the boot is not sealed, put in enough grease to push the old grease out. Then wipe off the old grease.

Task Completed ☐

8. Repeat this at all lubrication points and wipe all excess grease off the joints and fittings.

Task Completed ☐

Problems Encountered

_____

_____

_____

Instructor's Comments

_____

_____

_____

SUSPENSION AND STEERING JOB SHEET 34

Remove and Service a Strut and Coil Spring in a Rear Suspension

Name _____ Station _____ Date _____

NATEF Correlation

This Job Sheet addresses the following NATEF tasks:

C.2.1. Remove, inspect, and install coil springs and spring insulators.

C.2.4. Remove, inspect, and install strut cartridge or assembly, strut coil spring, and insulators (silencers).

Objective

Upon completion of this job sheet, you will be able to remove, inspect, and install coil springs and spring insulators, a strut cartridge, and the strut coil springs and insulators.

Tools and Materials

Floor jack

Spring compressing tool

Lift

Safety stands

Wooden block

Ball joint spreader tool

Hand tools

Protective Clothing

Goggles or safety glasses with side shields

Describe the vehicle being worked on:

Year _____ Make _____ Model _____

VIN _____ Engine type and size _____

PROCEDURE

1. Remove the rear seat and the package trim tray.   Task Completed ☐

2. Remove the wheel cover and loosen the wheel nuts.   Task Completed ☐

3. Lift the vehicle with a floor jack, and lower the chassis onto safety stands   Task Completed ☐
   so the rear suspension is allowed to drop downward.

4. Place a wooden block between the floor jack and the rear spindle on the   Task Completed ☐
   side where the strut and spring removal is taking place. Raise the floor jack
   to support some of the suspension system weight.

5. Remove the rear wheel, disconnect the nut from the small spring in the   Task Completed ☐
   lower arm, and remove the brake hose and antilock brake system (ABS)
   wire from the strut.

6. Remove the stabilizer bar link from the strut, and loosen the strut-to-spindle mounting bolts.

Task Completed ☐

7. Remove the upper support nuts under the package tray trim, and lower the floor jack to remove the strut from the knuckle. Remove the strut from the chassis.

Task Completed ☐

8. Following the manufacturer's recommended procedures, install a spring compressor on the coil spring, and tighten the spring compressor until all the spring tension is removed from the upper support.

Task Completed ☐

9. Install a bolt in the upper strut-to-spindle bolt hole and tighten two nuts on the end of this bolt. One nut must be on each side of the strut bracket.

Task Completed ☐

10. Clamp this bolt in a vise to hold the strut, coil spring, and spring compressor assembly.

Task Completed ☐

11. Using the special tool to hold the spring and strut from turning, loosen the nut from the strut rod nut.

Task Completed ☐

12. Remove the strut rod nut, upper support, and upper insulator.

Task Completed ☐

13. Remove the strut from the lower end of the spring.

Task Completed ☐

14. If the spring is to be replaced, rotate the compressor bolt until all the spring tension is removed from the compressing tool, then remove the spring from the tool.

Task Completed ☐

15. Inspect the lower insulator and spring seat on the strut. Is the spring seat warped or damaged? Is the insulator damaged in any way?

_____

_____

16. Describe the general condition of the strut.

_____

_____

17. List all the components that require replacement and explain the reasons for your diagnosis.

_____

_____

_____

_____

18. Visually inspect the coil spring, upper mount, insulator, and spring bumper. If any of these components are damaged, worn, or distorted, replacement is necessary. Describe the condition of each.

_____

_____

19. List all the components that require replacement and explain the reasons
    for your diagnosis.

    _____

    _____

    _____

    _____

Problems Encountered

_____

_____

_____

Instructor's Comments

_____

_____

_____

SUSPENSION AND STEERING JOB SHEET 35

Servicing Rear Suspensions and Shock Absorbers

Name _____ Station _____ Date _____

NATEF Correlation

This Job Sheet addresses the following NATEF tasks:

C.2.2. Remove, inspect, and install transverse links, control arms, bushings, and mounts.

C.3.1. Inspect, remove, and replace shock absorbers.

Objective

Upon completion of this job sheet, you will be able to inspect, remove, and install shock absorbers, transverse links, control arms, bushings, and mounts.

Tools and Materials

Lift   Control arm removing tool

Safety stands   Transmission jack

Hydraulic jack   Ball joint removal and replacement tools

Floor jack   Basic hand tools

Protective Clothing

Goggles or safety glasses with side shields

Describe the vehicle being worked on:

Year _____ Make _____ Model _____

VIN _____ Engine type and size _____

PROCEDURE

1. Lift the vehicle on a hoist with the chassis supported in the hoist and the control arms dropped downward. The vehicle may be lifted with a floor jack and the chassis supported on safety stands.

   Task Completed ☐

2. Remove the wheel and tire assembly.

   Task Completed ☐

3. Check the rebound bumpers on the control arms or chassis. If the rebound bumpers are severely worn, the shock absorbers may be worn out. What were the results of your inspection?

   _____

   _____

4. Check the mounting bolts and bushings for wear, looseness, or damage. If these components are loose, rattling noise will be evident and replacement of the bushings and bolts is necessary. What were the results of your inspection?

_____

_____

5. Check the shock absorbers and struts for oil leakage. A slight oil film on the lower oil chamber is acceptable. Any indication of oil dripping is not acceptable, and unit replacement is necessary. What were the results of your inspection?

_____

_____

6. Check the struts and shock absorbers for physical damage such as bends and severe dents or punctures. What were the results of your inspection?

_____

_____

7. Check the action of the shock absorbers by conducting a bounce test. Push the vehicle's bumper down with considerable weight applied on each corner of the vehicle. Release the bumper after pushing it down and observe the movement of the vehicle. How many upward bounces did the vehicle make before it settled?

_____

_____

8. A shock absorber can also be checked manually by disconnecting the lower end of the shock and moving the shock up and down as rapidly as possible. A good shock absorber should offer a strong, steady resistance to movement throughout the compression and rebound strokes. If a loss of resistance is experienced during either stroke, shock replacement is essential. What were the results of your inspection?

_____

_____

9. When front shock absorbers are being replaced, lift the front of the vehicle with a floor jack and support it with safety stands.

Task Completed ☐

10. When rear axle shock absorber replacement is necessary, raise the vehicle on a lift and support the rear axle with safety stands to prevent the shock absorbers from extending fully.

Task Completed ☐

11. Remove the upper shock mounting nut(s) or bolt(s) and the grommet.

Task Completed ☐

12. Remove the lower shock mounting nut(s) or bolt(s).

Task Completed ☐

13. Remove the shock absorber.

Task Completed ☐

14. Install the new shock absorber and grommet.

Task Completed ☐

15. Install and tighten the lower shock mounting nut(s) or bolt(s).   Task Completed ☐

16. Install and tighten the upper shock mounting nut(s) or bolt(s).   Task Completed ☐

17. Remove the stabilizer bar from the knuckle bracket.   Task Completed ☐

18. Remove the parking brake cable retaining clip from the lower control arm.   Task Completed ☐

19. If the car has electronic level control (ELC), disconnect the height sensor link from the control arm.   Task Completed ☐

20. Install a special tool to support the lower control arm in the bushing areas.   Task Completed ☐

21. Place a transmission jack under the special tool and raise the jack enough to remove the tension from the control arm bushing retaining bolts. If the car was lifted with a floor jack and supported on safety stands, place a floor jack under the special tool.   Task Completed ☐

22. Place a safety chain through the coil spring and around the lower control arm.   Task Completed ☐

23. Remove the bolt from the rear control arm bushing.   Task Completed ☐

24. Be sure the jack is raised enough to relieve the tension on the front bolt in the lower control arm, and remove this bolt.   Task Completed ☐

25. Lower the jack slowly and allow the control arm to pivot downward. When all the tension is released from the coil spring, remove the safety chain, coil spring, and insulators.   Task Completed ☐

26. Inspect the coil spring for distortion and proper free length. If the spring has a vinyl coating, check this coating for scratches or nicks. Check the spring insulators for cracks and wear.   Task Completed ☐

27. List all the components that require replacement, and explain the reasons for your diagnosis.

_____

_____

28. Remove the nut on the inner end of the suspension adjustment link, and disconnect this link from the lower control arm.   Task Completed ☐

29. Remove the cotter pin from the ball joint nut, and loosen, but do not remove, the nut from the ball joint stud.   Task Completed ☐

30. Use a special ball joint removal tool to loosen the ball joint in the knuckle.   Task Completed ☐

31. Remove the ball joint nut and the control arm.

32. Inspect the lower control arm for bends, distortion, and worn bushings.   Task Completed ☐

33. List the control arm and related parts that require replacement, and explain the reasons for your diagnosis.   Task Completed ☐

_____

_____

34. Use a special ball joint pressing tool to press the ball joint from the lower control arm.

Task Completed ☐

35. Use the same pressing tool with different adapters to press the new ball joint into the control arm.

Task Completed ☐

36. Install the ball joint stud in the knuckle and install the nut on the ball joint stud.

Task Completed ☐

37. Tighten the ball joint nut to the specified torque, and then tighten the nut an additional 2/3 turn. If necessary, tighten the nut slightly to align the nut castellations with the cotter pin hole in the ball joint stud, and install the cotter pin. What are the specifications for ball joint stud nut torque?

_____

_____

38. Snap the upper insulator on the coil spring. Install the lower spring insulator and the spring in the lower control arm.

Task Completed ☐

39. Make sure the top of the coil spring is properly positioned in relation to the front of the vehicle.

Task Completed ☐

40. Install the special tool on the inner ends of the control arm, and place the transmission jack or floor jack under the special tool.

Task Completed ☐

41. Slowly raise the transmission jack until the control arm bushing openings are aligned with the openings in the chassis.

Task Completed ☐

42. Install the bolts and nuts in the inner ends of the control arm. Do not torque these bolts and nuts at this time.

Task Completed ☐

43. Install the stabilizer-bar-to-knuckle bracket fasteners to the specified torque. What are the specifications?

_____

_____

44. Install the parking brake retaining clip.

Task Completed ☐

45. If the vehicle has ELC, install the height sensor link, and tighten the fastener to the specified torque. What are the specifications?

_____

_____

46. Install the suspension adjustment link and tighten the fastener to the specified torque. Install cotter pins as required. What are the specifications?

_____

_____

47. Remove the transmission jack or floor jack, and install the wheel-and-tire assembly.

Task Completed ☐

48. Lower the vehicle so that the vehicle's weight is on the tires and tighten the shock absorber mounting nut(s) or bolt(s) to specifications. What are the specifications?

_____

_____

49. Tighten the wheel hub nuts and lower control arm bolts and nuts to the specified torque. What are the specifications?

_____

_____

Problems Encountered

_____

_____

_____

Instructor's Comments

_____

_____

_____

SUSPENSION AND STEERING JOB SHEET 36

Rear Leaf Spring Diagnosis and Replacement

Name _____ Station _____ Date _____

NATEF Correlation

This Job Sheet addresses the following NATEF task:

C.2.3. Remove, inspect, and install leaf springs, leaf spring insulators (silencers), shackles, brackets, bushings, and mounts.

Objective

Upon completion of this job sheet, you will be able to remove, inspect, and install leaf springs, leaf spring insulators (silencers), shackles, brackets, bushings, and mounts.

Tools and Materials

Floor jack   Hand tools

Safety stands   Torque wrench

Protective Clothing

Goggles or safety glasses with side shields

Describe the vehicle being worked on:

Year _____ Make _____ Model _____

VIN _____ Engine type and size _____

PROCEDURE

1. Do the leaf springs squeak as the vehicle goes over a bump? _____ If   Task Completed ☐
   so, the leaf spring insulators are undoubtedly bad.

2. To replace the insulators, lift the vehicle with a floor jack and support the   Task Completed ☐
   frame on safety stands so the rear suspension moves downward.

3. With the vehicle weight no longer applied to the springs, the leaf spring   Task Completed ☐
   may be pried apart with a pry bar to remove and replace the silencers.

4. Is there a rattle coming from the area around the springs? _____ If so,   Task Completed ☐
   the shackle bushings, brackets, and mounts may be worn.

5. To check for worn shackles, shackle bushings, and brackets or mounts,   Task Completed ☐
   allow for the normal weight of the vehicle to rest on the springs.

6. Insert a pry bar between the rear outer end of the spring and the frame.
   Apply downward pressure on the bar and observe the rear shackle for
   movement. Shackle bushings, brackets, or mounts must be replaced if there
   is movement in the shackle. What did you observe?

   _____

   _____

7. The same procedure may be followed to check the front bushing in the main leaf. What did you observe?

_____

_____

8. Check the condition of the center bolt in both spring assemblies. Describe what you found. A broken spring center bolt may allow the rear axle assembly to move rearward on one side.

_____

_____

9. Check the curb riding height and compare your measurements to the specifications. How did they compare and what does this suggest?

_____

_____

10. If you found the springs to be sagging or if they must be replaced, prepare to replace them by raising the vehicle with a floor jack and placing safety stands under the frame.   Task Completed ☐

11. Lower the vehicle weight onto the safety stands, and leave the floor jack under the differential housing to support the rear suspension weight.   Task Completed ☐

12. Remove the nuts from the spring U-bolts, and remove the U-bolts and lower spring plate. The spring plate may be left on the rear shock absorber and moved out of the way.   Task Completed ☐

13. Be sure the floor jack is lowered sufficiently to relieve the vehicle weight from the rear springs.   Task Completed ☐

14. Remove the rear shackle nuts, plate, shackle, and bushings.   Task Completed ☐

15. Remove the front spring mounting bolt and remove the spring from the chassis. Check the spring center bolt to be sure it is not broken.   Task Completed ☐

16. Check the front hanger, bushing, and bolt; replace as necessary.   Task Completed ☐

17. Check the rear shackle, bushings, plate, and mount; replace the worn components.   Task Completed ☐

18. Reverse steps for removal to install the spring. Tighten all bolts and nuts to the specified torque.   Task Completed ☐

Problems Encountered

_____

_____

_____

Instructor's Comments

_____

_____

_____

SUSPENSION AND STEERING JOB SHEET 37

Electronic Air Suspension System Diagnosis

Name _____ Station _____ Date _____

NATEF Correlation

This Job Sheet addresses the following NATEF task:

C.3.3. Test and diagnose components of electronically controlled suspension systems using a scan tool; determine necessary action.

Objective

Upon completion of this job sheet, you will be able to diagnose, inspect, adjust, repair, or replace components of electronically controlled suspension systems.

Tools and Materials

Basic hand tools

Protective Clothing

Goggles or safety glasses with side shields

Describe the vehicle being worked on:

Year _____ Make _____ Model _____

VIN _____ Engine type and size _____

PROCEDURE

1. Be sure the air suspension system switch is turned on.   Task Completed ☐

2. Turn the ignition switch on for 5 seconds and then turn it off. Leave the driver's door open and the other doors closed.   Task Completed ☐

3. Ground the diagnostic lead located near the control module and close the driver's door with the window down.   Task Completed ☐

4. Turn the ignition switch on. The warning lamp should blink continuously at 1.8 times per second to indicate that the system is in the diagnostic mode. If the vehicle did not respond this way, describe how it did respond.

   _____

   _____

5. Perform test 1. Open and close the driver's door. The rear suspension should be raised for 30 seconds, lowered for 30 seconds, and raised for 30 seconds. If the vehicle did not respond this way, describe how it did respond. Did the air suspension lamp illuminate or blink?

   _____

   _____

6. Perform test 2. Open and close the driver's door. The right front suspension should be raised for 30 seconds, lowered for 30 seconds, and raised for 30 seconds. If the vehicle did not respond this way, describe how it did respond. Did the air suspension lamp illuminate or blink?

_____

_____

7. Perform test 3. Open and close the driver's door. The left front suspension should be raised for 30 seconds, lowered for 30 seconds, and raised for 30 seconds. If the vehicle did not respond this way, describe how it did respond. Did the air suspension lamp illuminate or blink?

_____

_____

8. Perform test 4. Open and close the driver's door. The compressor is cycled on and off at 0.25 cycles per second. This action is limited to 50 cycles. If the vehicle did not respond this way, describe how it did respond. Did the air suspension lamp illuminate or blink?

_____

_____

9. Perform test 5. Open and close the driver's door. The vent solenoid opens and closes at 1 cycle per second. If the vehicle did not respond this way, describe how it did respond. Did the air suspension lamp illuminate or blink during this sequence?

_____

_____

10. Perform test 6. Open and close the driver's door. The left front air valve opens and closes at 1 cycle per second, and the vent solenoid is opened. When this occurs, the left front corner of the vehicle should drop slowly. If the vehicle did not respond this way, describe how it did respond. Did the air suspension lamp illuminate or blink during this sequence?

_____

_____

11. Perform test 7. Open and close the driver's door. The right front air valve opens and closes at 1 cycle per second, and the vent solenoid is opened. This action causes the right front corner of the vehicle to drop slowly. If the vehicle did not respond this way, describe how it did respond. Did the air suspension lamp illuminate or blink during this sequence?

_____

_____

12. Perform test 8. Open and close the driver's door. During this test the right rear air valve opens and closes at 1 cycle per second and the vent valve is opened. This action should cause the right rear corner of the vehicle to drop slowly. If the vehicle did not respond this way, describe how it did respond. Did the air suspension lamp illuminate or blink during this sequence?

_____

_____

13. Perform test 9. Open and close the driver's door. The left rear solenoid opens and closes at 1 cycle per second and the vent valve is opened, which should cause the left rear corner of the vehicle to drop slowly. If the vehicle did not respond this way, describe how it did respond. Did the air suspension lamp illuminate or blink during this sequence?

_____

_____

14. Perform test 10. Return the module from the diagnostic mode to normal operation by disconnecting the diagnostic lead from ground, turning the ignition switch off, and depressing the brake pedal. Did the system return to the normal mode?

_____

_____

15. List all air suspension problems identified during the test sequence.

_____

_____

Problems Encountered

_____

_____

_____

Instructor's Comments

_____

_____

_____

SUSPENSION AND STEERING JOB SHEET 38

Remove and Install Front Wheel Bearings on a RWD Vehicle

Name _____ Station _____ Date _____

NATEF Correlation

This Job Sheet addresses the following NATEF task:

C.3.2. Remove, inspect, and service or replace front and rear wheel bearings.

Objective

Upon completion of this job sheet, you will be able to remove, inspect, and service or replace front and rear wheel bearings.

Tools and Materials

Ball peen hammer

Bearing repacker, optional

Clean paper

Drift

Grease

Hoist or jack stands

Installation tools, such as arbor press,
   press-fitting tools with an outside
   diameter approximately 0.010″ smaller
   than bore size, soft striking mallet, and
   wood block and hammer—if no other
   tools are available

Lint-free cloth

New bearings

New seals

Pliers or screwdriver

Service manual

Torque wrench

Wire, if needed

Wrenches

Protective Clothing

Goggles or safety glasses with side shields

Describe the vehicle being worked on:

Year _____ Make _____ Model _____

VIN _____ Engine type and size _____

Describe general condition:

_____

PROCEDURE (REMOVAL)

1. Raise the front end of the vehicle on a hoist or safely support it on jack
   stands. Do not support the vehicle on only a bumper jack. Remove the hub
   cap or wheel cover.   Task Completed ☐

2. Use a wrench to remove the wheel lug nuts. If the vehicle is equipped with disc brakes, loosen and remove the brake caliper mounting bolts. Support the caliper while disconnected on the lower A-frame or suspended by a wire loop.

Task Completed ☐
Not Applicable ☐

3. Use pliers or a screwdriver to remove the dust cover. Remove the cotter pin, and remove the adjusting nut.

Task Completed ☐

4. Jerk the rotor or drum assembly to loosen the washer and outer wheel bearing. If this step is not done easily, the drum brakes might have to be backed off.

Task Completed ☐

5. Remove the outer wheel bearing. Then pull the drum or rotor assembly straight off the spindle. Make sure the inner bearing or seal does not drag on the spindle threads.

Task Completed ☐

6. With the seal side down, lay the rotor or drum on the floor. Place the drive against the inner race of the bearing cone. Carefully tap out the old seal and inner bearing.

Task Completed ☐

WARNING: *Wear eye protection whenever using a hammer and drift punch.*

7. Record the part number of the seal on the Report Sheet for Removal and Installation of Front Wheel Bearings and Seals, to aid in selecting the correct placement. Discard the old seal.

Task Completed ☐

8. Clean and inspect the old bearing thoroughly.

Task Completed ☐

WARNING: *Do not use air pressure to spin the bearing during cleaning. A lack of lubrication can cause the bearing to explode, resulting in serious injury.*

9. Inspect the bearing to determine if it can be reused. Record the results on the Report Sheet for Removal and Installation of Front Wheel Bearings and Seals found at the end of this job sheet. If it must be replaced, record the part number on the report sheet.

Task Completed ☐

PROCEDURE (INSTALLATION)

1. Match the part numbers to make sure the new seal is correct for the application.

Task Completed ☐

2. By hand or with a bearing repacker, force grease through the cage and rollers or balls and on all surfaces of the bearing.

Task Completed ☐

3. Place the inner side of the drum or rotor face up. Use drivers to drive the new cup into the hub.

Task Completed ☐

4. Coat the hub cavity with the same wheel bearing grease to the depth of the bearing cup's smallest diameter. Apply a light coating of grease to the spindle.

Task Completed ☐

5. Place the inner bearing on the hub. Lightly coat the lip of the new seal with the same grease. Slide the seal onto the proper installation tool. Make sure the seal fits over the tool's adapter and the sealing lip points toward the bearing.

Task Completed ☐

6. Position the seal so it starts squarely in the hub without cocking. Tap the tool until it bottoms out. (When the sound of the striking mallet changes, the seal will be fully seated in the hub.)

Task Completed ☐

7. If the installation tool is unavailable, use a wood block and hammer to drive the seal. (Never hammer directly on seal.)

Task Completed ☐
Not Applicable ☐

8. Locate the lug nut and wheel bearing adjusting nut torque specifications in the service manual. Record the specifications on the Report Sheet for Removal and Installation of Front Wheel Bearings and Seals. Using these specifications and service manual procedures, install the hub assembly to the spindle.

Task Completed ☐

Problems Encountered

_____

_____

_____

Instructor's Comments

_____

_____

_____

Name _____ Station _____ Date _____

REPORT SHEET FOR REMOVAL AND INSTALLATION OF FRONT WHEEL BEARINGS AND SEALS		
Seal part number		
Bearing part number		
	Serviceable	*Nonserviceable*
Bearing inspection		
Cup		
Rollers		
Inner race		
Cage		
Overall condition		
Lug nut torque specification		
Bearing adjustment nut torque		
Initial torque		
Number of turns backed off		
Final torque		

Conclusions and Recommendations _____

_____

_____

SUSPENSION AND STEERING JOB SHEET 39

Road Test Vehicle and Diagnose Steering Operation

Name _____ Station _____ Date _____

NATEF Correlation

This Job Sheet addresses the following NATEF tasks:

D.1. Diagnose vehicle wander, drift, pull, hard steering, bump steer, memory steer, torque steer, and steering return concerns; determine necessary action.

Objective

Upon completion of this job sheet, you will be able to diagnose vehicle wander, drift, pull, hard steering, bump steering, memory steer, torque steer, and steering wheel return problems.

Tools and Materials

None required

Protective Clothing

Goggles or safety glasses with side shields

Describe the vehicle being worked on:

Year _____ Make _____ Model _____

VIN _____ Engine type and size _____

PROCEDURE

1. What is the customer's complaint recorded on this vehicle's repair order?

_____

_____

_____

Does the complaint present a possible safety concern for the operation of the vehicle? Consult with the instructor on your conclusion.

_____

_____

2. Check the tire pressure in each wheel and, if necessary, inflate to recommended pressures. Are there any obvious leaks?

_____

_____

3. Check each tire for damage of abnormal/excessive thread wear. Does any tire present a possible safety hazard during a road test?

_____

_____

4. Inspect the wheels and lug nuts. Is there any damage to any wheel? Do all the lug nuts appear securely tightened? If in doubt, use a torque wrench to check the torque on suspected lug nuts.

_____

_____

5. With engine off, rotate the steering wheel about 10 degrees side to side. Is there play or looseness in the steering? Perform the same check with engine operating at idle.

_____

_____

6. Clear the area rearward and to the side of the vehicle. If possible, have an assistant act as ground guide. Move the vehicle slowly while rotating the steering wheel from about midway of a full right turn to midway of a full left turn. Is the steering system firm with no looseness or binding?

_____

_____

7. If, for any reason, the vehicle is deemed unsafe to operate, consult the instructor for directions.

8. Road test vehicle using a variety of driving conditions from slow speed driving and cornering to normal cruising speed driving on a straight, level road surface. Check for the following abnormal steering conditions and note all abnormal behavior of the steering system.

   a. Vertical chassis oscillations—Did the vehicle's behavior seem normal? If not, why not?

   _____

   _____

   b. Chassis lateral waddle—Did the vehicle's behavior seem normal? If not, why not?

   _____

   _____

   c. Steering pull to right—Did the vehicle's behavior seem normal? If not, why not?

   _____

   _____

d. Steering pull to left—Did the vehicle's behavior seem normal? If not, why not?

_____

_____

e. Steering effort—Did the vehicle's behavior seem normal? If not, why not?

_____

_____

f. Tire squeal while cornering—Did the vehicle's behavior seem normal? If not, why not?

_____

_____

g. Bump steer—Did the vehicle's behavior seem normal? If not, why not?

_____

_____

h. Torque steer—Did the vehicle's behavior seem normal? If not, why not?

_____

_____

i. Memory steer—Did the vehicle's behavior seem normal? If not, why not?

_____

_____

j. Steering wheel return—Did the vehicle's behavior seem normal? If not, why not?

_____

_____

k. Steering wheel freeplay—Did the vehicle's behavior seem normal? If not, why not?

_____

_____

9. Return the vehicle to the shop and inspect suspension and steering to determine the cause of all abnormal conditions. List the necessary repairs and/or adjustments to correct all abnormal conditions that occurred during the road test.

_____

_____

_____

_____

Problems Encountered

_____

_____

_____

Instructor's Comments

_____

_____

_____

SUSPENSION AND STEERING JOB SHEET 40

Diagnosing Steering and Suspension Problems

Name _____ Station _____ Date _____

NATEF Correlation

This Job Sheet addresses the following NATEF task:

> **D.1.** Diagnose vehicle wander, drift, pull, hard steering, bump steer, memory steer, torque steer,
> and steering return concerns; determine necessary action.

Objective

Upon completion of this job sheet, you will be able to correctly diagnose problems in the suspension
and steering systems.

Tools and Materials

Service manual

Protective Clothing

Goggles or safety glasses with side shields

Describe the vehicle being worked on:

Year _____ Make _____ Model _____

VIN _____ Engine type and size _____

PROCEDURE

1. Carefully check the wear and inflation of each of the vehicle's tires. Describe
   their condition.

   _____

   _____

   _____

   _____

2. Explain what is indicated by the wear of the tires.

   _____

   _____

3. Take the vehicle on a road test. Check for the following problems and summarize when these problems occur and what is a likely cause of the problem.

    a. Excessive vertical chassis oscillations

    _____

    _____

    b. Chassis waddle

    _____

    _____

    c. Steering wander, pull, or drift

    _____

    _____

    d. High steering effort and/or binding

    _____

    _____

    e. Tire squeal while cornering

    _____

    _____

    f. Bump steer

    _____

    _____

    g. Torque steer

    _____

    _____

    h. Memory steer

    _____

    _____

    i. Poor steering wheel return

    _____

    _____

4. Define caster.

    _____

    _____

5. Define camber.

   _____

   _____

6. Define toe.

   _____

   _____

7. Typically, if the caster of the wheels is not within specifications, what symptoms will the vehicle have?

   _____

   _____

   _____

   _____

8. What effect does caster have on tire wear?

   _____

   _____

9. Typically, if the camber of the wheels is not within specifications, what symptoms will the vehicle have?

   _____

   _____

   _____

   _____

10. What effect does camber have on tire wear?

    _____

    _____

11. Typically, if the toe of the wheels is not within specifications, what symptoms will the vehicle have?

    _____

    _____

    _____

    _____

12. What effect does toe have on tire wear?

    _____

    _____

Problems Encountered

_____

_____

_____

Instructor's Comments

_____

_____

_____

SUSPENSION AND STEERING JOB SHEET 41

Prealignment Inspection

Name _____ Station _____ Date _____

NATEF Correlation

This Job Sheet addresses the following NATEF task:

D.2. Perform prealignment inspection; perform necessary action.

Objective

Upon completion of this job sheet, you will be able to conduct a complete prealignment check and inspection and determine necessary repairs prior to aligning the wheels of the vehicle.

Tools and Materials

Service manual

Dial indicator

Lift

Protective Clothing

Goggles or safety glasses with side shields

Describe the vehicle being worked on:

Year _____ Make _____ Model _____

VIN _____ Engine type and size _____

PROCEDURE

1. Road test the vehicle and drive the vehicle under the conditions cited by the customer as to when the complaint occurred.

   Task Completed ☐

2. Listen for any unusual noises related to the suspension and steering systems. Test the vehicle using a variety of driving conditions from slow speed driving and cornering to normal cruising speed driving on a straight, level road surface. Pay close attention to signs of steering pull, steering wander, erratic steering on road irregularities, high steering effort and binding, bump steer, torque steer, memory steer, steering wheel return problems, or the steering wheel not being centered while driving straight ahead. Describe the vehicle's behavior.

   _____

   _____

3. Check for excessive mud adhered to the chassis. Describe what you found and explain why this may affect wheel alignment.

_____

_____

4. Inflate the tires to the recommended pressure and note any abnormal tread wear or damage on each tire. What is the recommended air inflation and what did you find when inspecting the tires?

_____

_____

5. Check to make sure all of tires are the same size. Were they?

_____

_____

6. Check the front tires and wheels for radial runout. What did you find?

_____

_____

7. Check the suspension ride height. What did you find and what should you do?

_____

_____

8. Check the shock absorbers or struts for loose mounting bushings and bolts. Examine each shock absorber or strut for leakage. What did you find and what should you do?

_____

_____

9. Check the condition of each shock absorber or strut with a bounce test at each corner of the vehicle. What did you find and what should you do?

_____

_____

10. Raise the vehicle with the suspension supported and check the front wheel bearings for lateral movement. What did you find and what should you do?

_____

_____

11. Measure the ball joint radial and axial movement. What did you find and what should you do?

_____

_____

12. Inspect the control arms for damage and check the control arm bushings for wear. What did you find and what should you do?

_____

_____

13. Check all the steering linkages and tie-rod ends for looseness. What did you find and what should you do?

_____

_____

14. Check for worn stabilizer mounting links and bushings. What did you find and what should you do?

_____

_____

15. Check for loose steering gear mounting bolts and worn mounting brackets and bushings. What did you find and what should you do?

_____

_____

16. In your opinion, is this vehicle ready for an accurate four-wheel alignment?

_____

_____

Problems Encountered

_____

_____

_____

Instructor's Comments

_____

_____

_____

SUSPENSION AND STEERING JOB SHEET 42

Measuring Front and Rear Curb Riding Height

Name _____ Station _____ Date _____

NATEF Correlation

This Job Sheet addresses the following NATEF task:

D.3. Measure vehicle riding height; determine necessary action.

Objective

Upon completion of this job sheet, you will be able to measure front and rear curb riding height.

Tools and Materials

Tire pressure gauge

Service manual

Machinist rule or tape measure

Protective Clothing

Goggles or safety glasses with side shields

Describe the vehicle being worked on:

Year _____ Make _____ Model _____

VIN _____ Engine type and size _____

PROCEDURE

1. Check the trunk for extra weight. What is in the trunk? Should it be removed?

   _____

   _____

2. What is the recommended tire air pressure?

   _____

3. Check the tires for normal inflation pressure. Correct the air pressure if necessary. What were your findings?

   _____

4. Park the car on a level shop floor or alignment rack.   Task Completed ☐

5. Find the vehicle manufacturer's specified curb riding height measurement locations in the service manual. Record the specifications here:

_____

_____

_____

6. Measure and record the right front curb riding height. Your measurement was: _____

7. Measure and record the left front curb riding height. Your measurement was: _____

8. Measure and record the right rear curb riding height. Your measurement was: _____

9. Measure and record the left rear curb riding height. Your measurement was:

_____

10. Compare the measurement results to the specified curb riding height in the service manual. What do you conclude?

_____

_____

_____

_____

11. What should be done to the vehicle?

_____

_____

_____

_____

Problems Encountered

_____

_____

_____

Instructor's Comments

_____

_____

_____

SUSPENSION AND STEERING JOB SHEET 43

Measure Front and Rear Wheel Alignment Angles with a Computer-Based Wheel Aligner

Name _____ Station _____ Date _____

NATEF Correlation

This Job Sheet addresses the following NATEF tasks:

D.4.   Check and adjust front and rear wheel camber; perform necessary action.

D.5.   Check and adjust caster; perform necessary action.

D.6.   Check and adjust front wheel toe and center steering wheel.

D.7.   Check toe out on turns (turning radius); determine necessary action.

D.8.   Check SAI (steering axis inclination) and included angle; determine necessary action.

D.9.   Check and adjust rear wheel toe.

D.10. Check rear wheel thrust angle; determine necessary action.

D.11. Check for front wheel setback; determine necessary action.

Objective

Upon completion of this job sheet, you will be able to check and adjust the front and rear wheel camber, caster, front wheel toe, toe out on turns (turning radius) and rear wheel toe. You will also be able to check the steering axis inclination (SAI) and included angle, as well as rear wheel thrust angle and front wheel setback. After the alignment has been completed, you will also be able to center the steering wheel.

Tools and Materials

Computer-based alignment machine   Miscellaneous adjustments tools and shims

Hand tools   Tie-rod sleeve tool

Service manual

Protective Clothing

Goggles or safety glasses with side shields

Describe the vehicle being worked on:

Year _____ Make _____ Model _____

VIN _____ Engine type and size _____

Describe general operating condition:

_____

Describe the type and model of alignment machine being used:

_____

PROCEDURE

1. Lock the front turntables and drive the vehicle onto the alignment rack.   Task Completed ☐

2. Position the front wheels on the turntables.   Task Completed ☐

3. Make sure the rear wheels are properly positioned on the slip plates.   Task Completed ☐

4. Install the rim clamps and wheel sensors.   Task Completed ☐

5. Perform wheel sensor leveling and wheel runout compensation procedures.   Task Completed ☐

6. Select the specifications for the vehicle on the computer.   Task Completed ☐

7. Perform a prealignment inspection and record your findings and recommendations:

   _____

   _____

   _____

   _____

   _____

8. Measure ride height and compare to specifications.

   Left front—specified _____ measured _____

   Left rear—specified _____ measured _____

   Right front—specified _____ measured _____

   Right rear—specified _____ measured _____

9. Describe what can be done to correct ride height:

   _____

   _____

   _____

10. Measure front and rear suspension alignment angles following the prompts on the alignment machine's screen. Record the specifications and your measurements:

    Specified front camber _____

      Measured left front camber _____

      Measured right front camber _____

    Specified cross camber _____

      Measured cross front camber _____

    Specified front caster _____

      Measured left front caster _____

      Measured right front caster _____

Specified cross caster _____

    Measured cross front caster _____

Specified front SAI _____

    Measured left front SAI_____

    Measured right front SAI _____

    Measured included angle _____

Specified thrust angle _____

    Measured thrust angle _____

Specified front toe _____

    Measured left front toe _____

    Measured right front toe _____

    Measured total front toe _____

Specified rear camber_____

    Measured left rear camber _____

    Measured right rear camber _____

Specified rear toe _____

    Measured left rear toe _____

    Measured right rear toe_____

    Measured total rear toe _____

11. Measure the turning radius and compare to specifications. The specified turning radius for this vehicle is:_____

For a left turn, the turning radius of the right front wheel was: _____

For a left turn, the turning radius of the left front wheel was: _____

For a right turn, the turning radius of the right front wheel was:_____

For a right turn, the turning radius of the left front wheel was: _____

12. What are your conclusions from this test?

_____

_____

_____

13. State the necessary adjustments and repairs required to correct front and rear suspension alignment angles and justify your reason for each:

_____

_____

_____

_____

_____

_____

_____

14. List the suspension and steering service procedures that may require steering wheel centering after these procedures are completed.

_____

_____

_____

15. The steering wheel is centered properly with the front wheels straight ahead in the shop, but the steering wheel is not centered when driving the vehicle straight ahead. List the causes of this problem.

_____

_____

_____

_____

16. Lift the front end of the vehicle with a hydraulic jack and position safety stands under the lower control arms. Lower the vehicle onto the safety stands and place the front wheels in the straight-ahead position.   Task Completed ☐

17. Use a piece of chalk to mark each tie-rod sleeve in relation to the tie rod, and loosen the sleeve clamps.   Task Completed ☐

18. Position the steering wheel spoke in the position it was in while driving straight ahead during the road test. Turn the steering wheel to the centered position and note the direction of the front wheels. Which direction are the front wheels turned with the steering wheel centered?

_____

_____

19. If the steering wheel spoke is low on the left side while driving the vehicle straight ahead, use a tie-rod sleeve rotating tool to shorten the left tie rod and lengthen the right tie rod. A one-quarter turn on a tie-rod sleeve moves the steering wheel position approximately 1 inch. Turn the tie-rod sleeves the proper amount to bring the steering wheel to the centered position. If the steering wheel spoke is low on the right side while driving the vehicle straight ahead, lengthen the left tie rod and shorten the right tie rod. What did you need to do?

_____

_____

_____

_____

20. Mark each tie rod sleeve in its new position in relation to the tie rod.   Task Completed ☐

21. Tighten the clamp bolts to the specified torque. What is the spec?

_____

22. Lift the front chassis with a floor jack and remove the safety stands. Lower the vehicle onto the shop floor and check the steering wheel position during a road test. Were you able to center the steering wheel?

_____

_____

Problems Encountered

_____

_____

_____

Instructor's Comments

_____

_____

_____

SUSPENSION AND STEERING JOB SHEET 44

Inspect and Measure a Front Cradle

Name _____ Station _____ Date _____

NATEF Correlation

This Job Sheet addresses the following NATEF task:

D.12. Check front cradle (subframe) alignment; determine necessary action.

Objective

Upon completion of this job sheet, you will be able to check the front cradle (subframe) alignment.

Tools and Materials

Tram gauge

Service manual

Protective Clothing

Goggles or safety glasses with side shields

Describe the vehicle being worked on:

Year _____ Make _____ Model _____

VIN _____ Engine type and size _____

PROCEDURE

1. Raise the vehicle on a lift using the specified lifting points.

   Task Completed ☐

2. Inspect the front cradle mounts for looseness, damage, oil soaking, wear, and deterioration. Summarize your findings.

   _____

   _____

3. Inspect the front cradle for visible bends and damage. Summarize your findings.

   _____

   _____

4. Do the front cradle aligning hole(s) properly align with the matching holes in the chassis? _____ What are your service recommendations?

   _____

   _____

5. Locate the dimensions and specifications in the service manual that need to be measured when checking the front cradle. List these here.

_____

_____

_____

_____

6. Use a tram gauge to complete all measurements across the width of the cradle, starting at the front of the cradle; record your measurements.

_____

_____

_____

_____

7. Use a tram gauge to complete all front-to-rear measurements on the cradle, starting on the left side of the cradle; record your measurements.

_____

_____

_____

_____

8. Use a tram gauge to complete all diagonal cradle measurements, starting at the left side of the cradle; record your measurements.

_____

_____

_____

_____

9. What are your conclusions about the cradle, and what services do you recommend?

_____

_____

_____

_____

Problems Encountered

_____

_____

_____

Instructor's Comments

_____

_____

_____

SUSPENSION AND STEERING JOB SHEET 45

Inspect Tires for Inflation and Wear

Name _____ Station _____ Date _____

NATEF Correlation

This Job Sheet addresses the following NATEF tasks:

E.1. Diagnose tire wear patterns; determine necessary action.

E.2. Inspect tires; check and adjust air pressure.

Objective

Upon completion of this job sheet, you will be able to diagnose tire wear patterns and inspect tires.

Tools and Materials
Tire pressure gauge
Compressed air

Protective Clothing
Goggles or safety glasses with side shields

Describe the vehicle being worked on:
Year _____ Make _____ Model _____

VIN _____ Engine type and size _____

Describe the type and size of the tires on the vehicle:

_____

PROCEDURE

1. Visually inspect each tire's sidewalls for blemishes or damage. Record your findings here.

   _____

   _____

2. Answer these questions about tread wear patterns:

   a. What is indicated by wear on the edges or shoulders of tires?

   _____

   _____

   b. What should be done to correct the problem?

   _____

   _____

c. What is indicated by wear in the center of tires?

_____

_____

d. What should be done to correct the problem?

_____

_____

e. What is indicated by cracked tire treads?

_____

_____

f. What should be done to correct the problem?

_____

_____

g. What is indicated by wear on one edge or side of tires?

_____

_____

h. What should be done to correct the problem?

_____

_____

i. What is indicated by feathered edges or shoulders on tires?

_____

_____

j. What should be done to correct the problem?

_____

_____

k. What is indicated by bald spots on tires?

_____

_____

l. What should be done to correct the problem?

_____

_____

3. Visually inspect each tire on the vehicle and give your service recommendations.

_____

_____

_____

4. Locate the recommended tire pressure of the tires on this vehicle. Where did you find those recommendations and what are they?

_____

_____

5. Using an air pressure gauge, check the pressure in each tire. Record your results.

_____

_____

6. Correct the air pressure, if necessary.   Task Completed ☐

Problems Encountered

_____

_____

_____

Instructor's Comments

_____

_____

_____

SUSPENSION AND STEERING JOB SHEET 46

Diagnosing Wheel and Tire Problems

Name _____ Station _____ Date _____

NATEF Correlation

This Job Sheet addresses the following NATEF task:

E.3. Diagnose wheel/tire vibration, shimmy, and noise; determine necessary action.

Objective

Upon completion of this job sheet, you will be able to diagnose wheel and tire vibration, shimmy, and noise problems.

Tools and Materials

Tread wear gauge

Protective Clothing

Goggles or safety glasses with side shields

Describe the vehicle being worked on:

Year _____ Make _____ Model _____

VIN _____ Engine type and size _____

PROCEDURE

1. Drive the vehicle at a variety of speeds and under a variety of conditions and note all noises and handling problems that occurred during the road test.

   _____

   _____

   _____

2. Check the suspension, steering, driveline, and brake system to make sure they are not the cause of the problems. Describe the results of this check.

   _____

   _____

3. Check the wear patterns of the tires to help eliminate the suspension and steering systems as causes of the problem. Describe the wear pattern of the tires.

   _____

   _____

4. Check the wear of the tires with a tread wear gauge or by looking for tread bars on the tires. Describe the wear of each tire.

   _____

   _____

5. Check the inflation of all of the tires and compare the pressure to specifications. What did you read and what are the specifications?

   _____

   _____

6. If the results from the above all indicate that the tires; the suspension, steering, and brake systems; and the driveline are in good condition and yet a noise was evident during the test drive, suspect the tire design as the cause of the problem. Describe the tread design of the tires. Do all of the tires have the same design?

   _____

   _____

7. What can you conclude about the source of the abnormal noise?

   _____

   _____

8. Describe any vibration problems felt during the road test. If the tires felt as if they were bouncing along, the problem is typically called wheel tramp. If the tires seemed to vibrate while turning, the problem is called shimmy.

   _____

   _____

   _____

   _____

9. If wheel tramp is the problem, the tire and wheel assemblies need to be checked for static imbalance. If wheel shimmy is the problem, check the tire and wheel assemblies for dynamic imbalance. Describe the results of this check.

   _____

   _____

10. If the tire and wheel assembly are both statically and dynamically balanced, check the wheel hub or axle for excessive runout.   Task Completed ☐

11. If runout is not the problem, check for loose, worn, or damaged wheel bearings.   Task Completed ☐

Problems Encountered

_____

_____

_____

Instructor's Comments

_____

_____

_____

SUSPENSION AND STEERING JOB SHEET 47

Tire Rotation

Name _____ Station _____ Date _____

NATEF Correlation

This Job Sheet addresses the following NATEF task:

E.4. Rotate tires according to manufacturer's recommendations.

Objective

Upon completion of this job sheet, you will be able to rotate tires according to manufacturer's recommendations.

Tools and Materials

Torque wrench

Lift or jack

Safety stands

Protective Clothing

Goggles or safety glasses with side shields

Describe the vehicle being worked on:

Year _____ Make _____ Model _____

VIN _____ Tire type and size _____

PROCEDURE

1. Most car manufacturers recommend tire rotation at specified intervals to obtain maximum tire life. The exact tire rotation procedure depends on the model year, the type of tires, and whether the vehicle has a conventional spare or a compact spare. What is the recommended time interval for tire rotation on this vehicle?

   _____

   _____

2. Tire rotation procedures do not include the compact spare. The vehicle manufacturer provides tire rotation information in the owner's manual and service manual. Describe the procedure recommended for this vehicle.

   _____

   _____

3. Loosen the lug nuts for all four wheels.   Task Completed ☐

4. Raise the vehicle so all four wheels are suspended.   Task Completed ☐

5. Remove the lug nuts for one wheel, then remove the wheel.   Task Completed ☐

6. Move to the wheel where the "just removed" wheel should be located.   Task Completed ☐

7. Remove the lug nuts for that wheel, then remove the wheel.   Task Completed ☐

8. Install the wheel removed in step 5 to its new location.   Task Completed ☐

9. Move to the wheel where the "just removed" wheel should be located.   Task Completed ☐

10. Remove the lug nuts for that wheel, then remove the wheel.   Task Completed ☐

11. Install the wheel removed in step 7 to its new location.   Task Completed ☐

12. Move to the wheel where the "just removed" wheel should be located.   Task Completed ☐

13. Remove the lug nuts for that wheel, then remove the wheel.   Task Completed ☐

14. Install the wheel removed in step 10 to its new location.   Task Completed ☐

15. Move to the wheel where the "just removed" wheel should be located.   Task Completed ☐

16. Remove the lug nuts for that wheel, then remove the wheel.   Task Completed ☐

17. Install the wheel removed in step 13 to its new location.   Task Completed ☐

18. Lower the vehicle and set its weight onto the tires.   Task Completed ☐

19. Tighten all of the lug nuts to the torque specifications. What is that specification and what tightening pattern did you use for the lug nuts?

_____

_____

Problems Encountered

_____

_____

_____

Instructor's Comments

_____

_____

_____

SUSPENSION AND STEERING JOB SHEET 48

Tire and Wheel Runout Measurement

Name _____ Station _____ Date _____

NATEF Correlation

This Job Sheet addresses the following NATEF task:

E.5. Measure wheel, tire, axle, and hub runout; determine necessary action.

Objective

Upon completion of this job sheet, you will be able to measure wheel, tire, axle, and hub runout.

Tools and Materials

Basic hand tools

Dial indicator

Protective Clothing

Goggles or safety glasses with side shields

Describe the vehicle being worked on:

Year _____ Make _____ Model _____

VIN _____ Engine type and size _____

Wheel size and type _____ Tire size and manufacturer _____

PROCEDURE

1. Position a dial indicator against the center of the tire tread as the tire is rotated slowly to measure radial runout.

   Radial runout _____ Specified radial runout _____

2. Mark the highest point of radial runout on the tire with a crayon, and mark the valve stem position on the tire.   Task Completed ☐

3. If the radial tire runout is excessive, demount the tire and check the runout of the wheel rim with a dial indicator positioned against the lip of the rim while the rim is rotated.

   Wheel radial runout _____ Specified wheel radial runout _____

4. Use a crayon to mark the highest point of radial runout on the wheel rim.   Task Completed ☐

5. If the highest point of wheel radial runout coincides with the chalk mark from the highest point of maximum tire radial runout, the tire may be rotated 180 degrees on the wheel to reduce radial runout. Tires or wheels with excessive runout are usually replaced. Does the highest point of tire radial runout coincide with the highest point of wheel radial runout? State the required action to correct excessive radial runout and the reason for this action.

_____

_____

_____

_____

6. Position the dial indicator on the face of the wheel hub or axle flange. Rotate the hub or flange slowly to measure radial runout.

Radial runout _____ Specified radial runout _____

7. What can you conclude?

_____

_____

_____

8. Position a dial indicator located against the sidewall of the tire to measure lateral runout.

Tire lateral runout _____

Specified tire lateral runout _____

9. Chalk mark the tire and wheel at the highest point of radial runout.   Task Completed ☐

10. If the tire runout is excessive, the tire should be demounted from the wheel and the wheel lateral runout measured.   Task Completed ☐

11. If the tire runout was excessive, measure the wheel lateral runout with a dial indicator positioned against the edge of the wheel as the wheel is rotated.

Wheel lateral runout _____

Specified wheel lateral runout _____

12. State the required action to correct excessive wheel lateral runout and the reason for this action.

_____

_____

_____

_____

13. Position the dial indicator on the face of the wheel hub or axle flange. Rotate the hub or flange slowly to measure lateral runout.

   Lateral runout _____

   Specified lateral runout _____

14. What can you conclude?

   _____

   _____

   _____

   _____

Problems Encountered

_____

_____

_____

Instructor's Comments

_____

_____

_____

SUSPENSION AND STEERING JOB SHEET 49

Balance a Tire and Wheel Off the Vehicle

Name _____ Station _____ Date _____

NATEF Correlation

This Job Sheet addresses the following NATEF task:

E.7. Balance wheel and tire assembly (static and dynamic).

Objective

Upon completion of this job sheet, you will be able to remove a tire from a vehicle, computer balance the tire, and reinstall the tire on the vehicle.

Tools and Materials

Lug wrench

Torque wrench

Sockets

Service manual

Protective Clothing

Goggles or safety glasses with side shields

Describe the type of tire and wheel you are working on:

_____

What is the brand and model of tire balancer you are using?

_____

PROCEDURE

1. Raise and support the vehicle.   Task Completed ☐

2. Loosen the lug nuts and remove them.   Task Completed ☐

3. Remove the tire from the vehicle.   Task Completed ☐

4. Using the proper procedures, for the specific equipment, balance the tire   Task Completed ☐
   on your computerized tire balancer.

5. Reinstall the tire on the vehicle.   Task Completed ☐

6. Reinstall and torque lug nuts to the manufacturer's specifications.   Task Completed ☐

7. Lower the vehicle to the ground.   Task Completed ☐

Problems Encountered

_____

_____

_____

Instructor's Comments

_____

_____

_____

SUSPENSION AND STEERING JOB SHEET 50

Dismounting and Mounting a Tire on a Wheel Assembly

Name _____ Station _____ Date _____

NATEF Correlation

This Job Sheet addresses the following NATEF task:

E.8.   Dismount, inspect, and remount tire on wheel.

E.10.   Reinstall wheel; torque lug nuts.

Objective

Upon completion of this job sheet, you will be able to dismount, inspect, and remount a tire on a wheel and to reinstall the wheel and torque the lug nuts.

Tools and Materials

Breaker bar   Wheel chocks

Tire machine   Safety stands

Hydraulic floor jack

Protective Clothing

Goggles or safety glasses with side shields

Describe the vehicle being worked on:

Year _____ Make _____ Model _____

VIN _____ Tire size _____ Wheel diameter _____

PROCEDURE

1. Determine the proper lift points for the vehicle. Describe them here:

   _____

   _____

2. Place the pad of the hydraulic floor jack under one of the lift points for the left side of the front of the vehicle. Describe this location.

   _____

   _____

3. Raise the pad just enough to make contact with the vehicle.   Task Completed ☐

4. Make sure the vehicle's transmission is placed in park or in a low gear (if there is a manual transmission) and apply the parking brake.   Task Completed ☐

5. Place blocks or wheel chocks at the rear tires to prevent the vehicle from rolling.

Task Completed ☐

6. Remove any hub or bolt caps to expose the lug nuts or bolts.

Task Completed ☐

7. Refer to the service or owner's manual to determine the order the lugs ought to be loosened. If this is not available, make sure they are alternately loosened. Describe the order you will use to loosen and remove the lug nuts.

_____

_____

8. Get the correct size socket or wrench for the lug nuts. The size is: _____

9. Using a long breaker bar, loosen each of the lug nuts but do not remove them.

Task Completed ☐

10. Raise the vehicle with the jack. Make sure it is high enough to remove the wheel assembly. Once it is in position, place a safety stand under the vehicle. Describe the location of the stand.

_____

_____

11. Lower the vehicle onto the stand and lower and remove the jack from the vehicle.

Task Completed ☐

12. Remove the lug nuts and the tire and wheel assembly.

Task Completed ☐

13. Place the lug nuts in a place where they will not be kicked around or lost.

Task Completed ☐

14. Place the tire and wheel assembly close to the tire machine.

Task Completed ☐

15. Release the air from the tire, then remove the valve stem core.

Task Completed ☐

16. Set the machine to unseat the tire from its rim, and then unseat it.

Task Completed ☐

17. Once both sides of the tire are unseated, place the tire and wheel onto the machine. Then depress the pedal that clamps the wheel to the tire machine.

Task Completed ☐

18. Lower the machine's arm into position on the tire and wheel assembly.

Task Completed ☐

19. Insert the tire iron between the upper bead of the tire and wheel. Depress the pedal that causes the wheel to rotate. Do the same with the lower bead.

Task Completed ☐

20. After the tire is totally free from the rim, remove the tire.

Task Completed ☐

21. Prepare the wheel for the mounting of the new tire by using a wire brush to remove all dirt and rust from the sealing surface. Apply rubber compound to the bead area of the tire.

Task Completed ☐

22. Place the tire onto the wheel and lower the arm into place. As the machine rotates the wheel, the arm will force the tire over the rim.

Task Completed ☐

23. After the tire is completely over the rim, install the air ring over the tire. What does the air ring do?

   _____

   _____

24. Reinstall the valve stem core and inflate the tire to the recommended inflation. What is that inflation?

   _____

25. Move the assembly over to the vehicle.   Task Completed ☐

26. Place the assembly onto the hub and start each of the lug nuts.   Task Completed ☐

27. Hand tighten each lug nut so that the wheel is fully seated against the hub.   Task Completed ☐

28. Put the hydraulic jack into position and raise the vehicle enough to remove the safety stands.   Task Completed ☐

29. Lower the jack so the tire is seated on the shop floor.   Task Completed ☐

30. Following the reverse pattern as used to loosen the lug nuts, tighten the lug nuts according to specifications. The order you used to tighten was:

   _____

   _____

   The recommended torque spec is: _____

31. What happens if you overtighten the lug nuts?

   _____

   _____

   _____

32. Install the bolt or hub cover.   Task Completed ☐

Problems Encountered

_____

_____

_____

Instructor's Comments

_____

_____

_____

SUSPENSION AND STEERING JOB SHEET 51

Servicing Wheels and Tires Equipped with TPMS

Name _____ Station _____ Date _____

NATEF Correlation:

This Job Sheet addresses the following NATEF tasks:

E.9. Dismount, inspect, and remount tire on wheel equipped with tire pressure sensor.

E.13. Inspect, diagnose, and calibrate tire pressure monitoring system.

Objective

Upon completion of this job sheet, you will be able to service wheels and tires equipped with a tire pressure monitoring system (TPMS).

Tools and Materials

Service manual

Scan tool

Tire changer

Protective Clothing

Goggles or safety glasses with side shields

Describe the vehicle being worked on:

Year _____ Make _____ Model _____

VIN _____ Engine type and size _____

PROCEDURE

Tire and Sensor Removal

1. Raise the vehicle, and support it with safety stands in the proper locations.   Task Completed ☐

2. Remove the wheel with the faulty sensor.   Task Completed ☐

3. Remove the valve core and cap, and release air from the tire.   Task Completed ☐

4. After all air has been expelled from the tire, remove the nut, washer, and outer grommet holding the tire pressure monitor valve and drop the sensor inside the tire.   Task Completed ☐

5. Remove any balance weights, and then break the outside bead loose from the wheel with a tire changer. Position the wheel so the valve stem is 90 degrees from the bead breaker.   Task Completed ☐

   NOTE: *Be careful not to damage the tire pressure monitor valve due to interference between the tire pressure monitor valve and tire bead.*

6. Take out the tire pressure monitor valve from the tire and remove the bead on the lower side.

7. Write down the transmitter ID or color of the sensor. When tires and wheels or the tire pressure monitor is replaced, the transmitter ID must be programmed to the electronic control unit. What is the transmitter ID?

_____

   NOTE: *The transmitter ID is on the tire pressure monitor valve assembly and cannot be seen after it is installed in the tire and wheel. Make sure you make a record of the transmitter ID before installing it.*

8. Remove the inner grommet from the tire pressure monitor valve. Carefully inspect it. Make sure it is not cracked or damaged. If necessary, replace the grommet with a new washer and nut. What is the condition of the grommet?

_____

_____

9. When replacing the tire pressure monitor valve, replace it with the one that has the same identification number or paint color.   Task Completed ☐

10. Install the inner grommet to the tire pressure monitor valve.   Task Completed ☐

11. Inspect the tire pressure monitor valve for damage. If the valve is deformed or damaged, replace it. Describe its condition.

_____

_____

_____

12. After tire service, install the lower tire bead.   Task Completed ☐

13. Position the sensor of the tire pressure monitor valve in the wheel. Make sure it is facing the correct way. According to the service manual, how should the assembly be positioned?

_____

_____

14. Tighten the air pressure monitor valve to specifications. What is the required tightening torque?

_____

15. When installing the tire, take care not to damage the sensor by making sure that the monitor valve does not interfere with the tire bead.   Task Completed ☐

16. Install the upper bead. Make sure that the sensor is not clamped by the bead.   Task Completed ☐

17. After the tire has fully seated and filled with air, mount it on the vehicle.   Task Completed ☐

18. Register the tire monitor sensor with the control unit according to the pro-
    cedure outlined in the service manual. Briefly describe this procedure.

    _____

    _____

19. Remove the valve core to rapidly release air and check that the warning
    lamp is illuminated. If not, the system may be defective. What happened?

    _____

    _____

20. Refill the tire with the proper amount of pressure. What is the required
    inflation for this tire and vehicle?

    _____

21. If there is a malfunction in the tire pressure monitor system, the ECU blinks
    the tire pressure warning lamp. The result of this diagnosis is stored in the
    tire pressure monitor ECU. When the vehicle is running, does the lamp go
    out?

    _____

    _____

Diagnosis

1. Interview the customer to obtain a description of the conditions when the
   indicator came on. Find out if the customer checked and/or adjusted tire
   pressures since the indicator came on. What did you find?

   _____

   _____

2. Under what conditions will a tire indicator be illuminated?

   _____

   _____

   What non-system events can cause the indicators to illuminate?

   _____

   _____

   NOTE: *If a flat tire is replaced with the spare tire, and the flat tire is stored in the cargo area,
   the low pressure indicator will stay on but the appropriate tire indicator will go off. This
   prevents the customer from thinking there is a problem with the spare tire.*

3. Observe the indicators for the Tire Pressure Monitoring System in the
   instrument panel while you turn the ignition on. Describe what happened
   to the lights.

   _____

   _____

4. Based on this check, does it seem that the system is working properly?

_____

_____

5. If the vehicle is safe, take it for a road test and try to duplicate the conditions of the customer's concern. Did you observe any system problems?

_____

_____

6. If an indicator did not come on during the road test, check for loose terminals or damaged wires. What did you find?

_____

_____

7. Connect a scan tool to the vehicle and enter the Tire Pressure Monitoring System.   Task Completed ☐

8. Record all codes that were retrieved and state what is indicated by each.

_____

_____

9. After troubleshooting, clear the DTCs, and test-drive the vehicle. Make sure no indicators come on.   Task Completed ☐

Calibration

1. When a tire pressure sensor or ECU is replaced, the sensor ID must be memorized by the control unit. To make sure the control unit memorizes the correct ID, the vehicle with the new sensor must be at least 3 m (10 ft) from any other TPMS pressure sensor not installed on that vehicle.   Task Completed ☐

2. On many vehicles after rotating the tires or replacing a tire pressure sensor, drive the vehicle for at least 40 seconds at a speed of 15 mph (24 km/h) or more, and all the sensor IDs will be memorized automatically. Is this true of the vehicle you are working on?

_____

_____

3. When replacing the TPMS control unit, use a scan tool to memorize IDs.   Task Completed ☐

4. Describe the procedure for memorizing the IDs.

_____

_____

_____

_____

5. After the IDs are memorized, reduce the pressure in all four tires to less than the appropriate specification, and check to see that the four tire indicators come on. What happened?

_____

_____

6. Refill the tires to the proper inflation.   Task Completed ☐

Problems Encountered

_____

_____

_____

Instructor's Comments

_____

_____

_____

SUSPENSION AND STEERING JOB SHEET 52

Tire Service

Name _____ Station _____ Date _____

NATEF Correlation

This Job Sheet addresses the following NATEF task:

E.11. Inspect and repair tire and wheel assembly for air loss; perform necessary action.

E.12. Repair tire using internal patch.

Objective

Upon completion of this job sheet, you will be able to inspect and determine if the tire can be repaired, then repair it if possible.

Tools and Materials

Tire pressure gauge

Tank of water

Tire crayon or chalk

Tire plug kit

Cold patch tire repair kit

Protective Clothing

Goggles or safety glasses with side shields

Describe the vehicle being worked on:

Year _____ Make _____ Model _____

VIN _____ Tire size _____

PROCEDURE

1. Carefully inspect the tire.

   A. Do the wear indicators show?

   _____

   _____

   B. Are the belts or tire fabric exposed?

   _____

   _____

   C. Are there bulges or blisters on the tire's sidewalls?

   _____

   _____

D. Is there evidence of ply separation?

_____

_____

E. Are the beads of the tire cracked or broken?

_____

_____

F. Are there cracks or cuts anywhere on the tire?

_____

_____

2. Based on this inspection, can the tire be safely repaired?

_____

_____

3. If the tire is repairable, inflate it to its maximum allowable pressure. What is that pressure and where did you find this specification?

_____

_____

4. Submerge the tire into a tank of water.   Task Completed ☐

5. If bubbles appear in the water, locate and mark the location of the leak with the tire crayon or chalk.   Task Completed ☐

6. If no bubbles appeared, rotate the tire so a new area is submerged in the water.   Task Completed ☐

7. If bubbles now appear in the water, locate and mark the location of the leak with the tire crayon or chalk.   Task Completed ☐

8. Once the leak has been located, remove the tire from the water and mark the tire at the location of the valve stem so the tire can be reinstalled in the same position to maintain proper balance.   Task Completed ☐

9. With a tire machine, separate the tire from the wheel.   Task Completed ☐

10. Buff the area inside the tire around the puncture with a wire brush or wire buffing wheel.   Task Completed ☐

11. Choose a plug or patch to repair the tire. What will you use?

_____

_____

12. To install a plug:
    a. Select a plug that is slightly larger than the puncture and insert the plug into the eye of the insertion tool.   Task Completed ☐

    b. Wet the plug and the insertion tool with vulcanizing fluid.   Task Completed ☐

    c. While holding and stretching the plug, pull the plug into the puncture from the inside of the tire. The head of the plug should contact the inside of the tire.   Task Completed ☐

    d. Cut the plug off 1/32-inch from the tread surface. Be careful not to stretch the plug while cutting it.   Task Completed ☐

13. To install a cold patch:
    a. Apply vulcanizing fluid to the buffed area and allow it to dry until it is tacky.   Task Completed ☐

    b. Peel the backing from the patch and center the patch over the puncture.   Task Completed ☐

    c. Run a stitching tool back and forth over the patch to improve bonding.   Task Completed ☐

14. To install a hot patch:
    a. Apply vulcanizing fluid to the buffed area.   Task Completed ☐

    b. Peel the backing from the patch and install the patch so it is centered over the puncture on the inside of the tire.   Task Completed ☐

    c. Clamp the heating element over the patch and leave it there for the time recommended by the manufacturer.   Task Completed ☐

    d. After the heating element has been removed, allow the patch to cool for a few minutes.   Task Completed ☐

    e. Check the patch to make sure it is bonded tightly to the tire.   Task Completed ☐

15. Reinstall the tire to the wheel.   Task Completed ☐

16. Inflate the tire to the proper pressure.   Task Completed ☐

17. Check the repair by submerging that point of the tire into water.   Task Completed ☐

Problems Encountered

_____

_____

_____

Instructor's Comments

_____

_____

_____

— NOTES —

— NOTES —

— NOTES —